GUERRILLAS
IN THE
MIST

To Colonel Kevin A. Conry USMC, a straight shooter amid a sea of crooked barrels.

—*Semper Fidelis.*

GUERRILLAS IN THE MIST

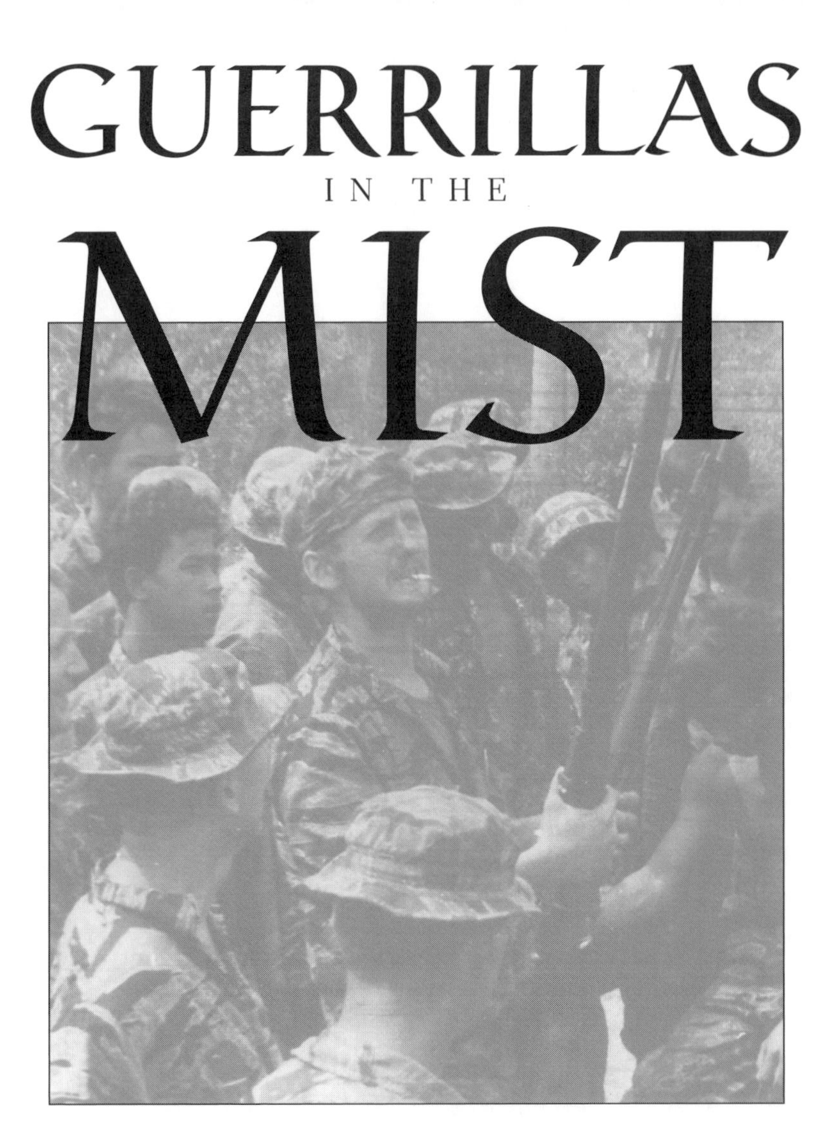

A Battlefield Guide to Clandestine Warfare

Bob Newman

Foreword by Robert K. Brown

Paladin Press • Boulder, Colorado

Also by Bob Newman:

The Ultimate Outdoorsman: Critical Skills for Traveling, Surviving, and Enjoying Your Time in the Wilderness (video)
Wilderness Wayfinding: How to Survive in the Wilderness as You Travel

Guerrillas in the Mist:
A Battlefield Guide to Clandestine Warfare
by Bob Newman
Foreword by Robert K. Brown

ISBN 10: 0-87364-944-3
ISBN 13: 978-0-87364-944-5
Printed in the United States of America

Published by Paladin Press, a division of
Paladin Enterprises, Inc.,
Gunbarrel Tech Center
7077 Winchester Circle
Boulder, Colorado 80301 USA
+1.303.443.7250

Direct inquiries and/or orders to the above address.

Contents

Warning

This book was written for elite-unit military personnel and students of military history. It was not intended for use by anyone whose aim or wish is to overthrow the United States government or any other government by force or any other means, nor was it intended for use as a demolitions manual. Conspiring, planning, or attempting to overthrow the U.S. government by force or other violent means, and the illegal use or misuse of explosives or booby traps, are serious crimes with harsh punishment for those convicted. The author, publisher, and distributors of this book disclaim any responsibility for the use or misuse of any information presented herein. *This book is for academic study only.*

Foreword

As I write this, I am on my way to yet another rebellion against tyranny, this one in long-suffering Albania, once the most xenophobic of hard-core Communist nations. After completing the Albania mission, I will head south to the Dark Continent to report on the troubles in Zaire. Thoughts of these, my most recent rebellions, as well as dozens of others I have seen, have caused me to consider the role of the guerrilla in the birth of countless nations—including America—as well as in the liberation of innumerable peoples from oppressive governments, military juntas, and tin-pot dictators. Still, most guerrilla-led rebellions fail miserably. So the questions that anyone who has an interest in rebellions (or expects to be involved in one) must ask are, why do they fail and how can a guerrilla movement succeed?

The answers to these questions are the whole point of this remarkable book, which is the first truly practical, hard-hitting manual on how to establish, equip, train, and successfully employ a guerrilla unit. Make no mistake about it, *Guerrillas in the Mist* is the "bible" of guerrilla warfare.

I first became associated with Gunnery Sergeant Bob Newman, USMC (Retired), when he wrote his first article for *Soldier of Fortune* back in 1991, shortly after his return from the Persian Gulf War, where he served with a highly decorated Marine infantry battalion as a platoon sergeant. (He is now *SOF*'s contributing editor for Gulf War Veterans Affairs.) His diverse background in the Corps, which includes tours in recon (airborne- and combat-diver qualified), as an instructor at the notorious Navy SERE School in Maine, Landing Force Training Command (Pacific), and the esteemed Staff Noncommissioned Officer Academy at Camp Geiger (where he was the chief warfighting instructor for the Advanced Course), and in both the Corps' Marine Amphibious Unit (MAU) and Marine Expeditionary Unit-Special Operations Capable (MEU-SOC) battle configurations, makes him uniquely qualified to write this book. As you may be aware, I choose the places where my written words land very carefully, and this, the latest of the gunny's books in a long, impressive list of them, is as secure an LZ as there is.

Read this book from start to finish, study it, listen to the gunny's words of wisdom, and you will stand a much improved chance of coming away from the fight with the smell of victory on your uniform.

—Lt. Col. Robert K. Brown, USAR (Ret.)
Publisher, *Soldier of Fortune* magazine

Acknowledgments

As is guerrilla warfare, writing a book is a joint effort. Through their insight, tactical expertise, comprehension of operational art, and esprit de corps, Marines like Colonel Kevin Conry, Colonel J.L. Clark, Lieutenant Colonel John Bass, Lieutenant Colonel J.W. Muth III, Major Keith Kelly, Major Bryan McCoy, 1st Sergeant Brad Delauter, 1st Sergeant George Misko, 1st Sergeant Rick Pelow, Master Sergeant Ron Wendt, Gunnery Sergeant Jeff Carothers, Gunnery Sergeant Joe Gill, and Gunnery Sergeant "Lou" Gregory, as well as professional sailors like Commander Bob Fant, Commander Tim Sullivan, and Master Chief Petty Officer Tom Keith, all had a hand in this work. Nor would any of this have been possible without the technical expertise and attention to detail of my editor, Karen Pochert; publisher, Peder Lund; editorial director, Jon Ford; proofer, Donna DuVall; art director, Fran Milner; art designer Barb Beasley; and the people whom writers often forget to recognize, without whom not a single book could be sold: the shipping crew (Tim Dyrendahl, Mike Kerr, Chris Kuhn, Ray Lyman, Susan Newman, Sash, and Dan Stone), sales and marketing directors (Wendy Apps and Tina Mills, respectively), video production manager (Mike Janich), chief financial officer (Dana Rogers), and the experts in the front office (Beverly Bayer,

Wanda Bennett, Tom Laidlaw, Cindy Nolting, Marilyn Ranson, and Paula Grano).

Thanks also go to Lieutenant Colonel Bob Brown for pressing on.

And, as always, my thanks to Susan.

"I have only come here seeking knowledge, things they would not teach me of in college."

—Sting

Introduction

When the cities of Hiroshima and Nagasaki went up in massive balls of fission, incinerating and fatally radiating hundreds of thousands of Japanese in little more than an instant, the world gasped in shock and wonder. Military strategists did too, and many became immediate conscripts to the belief that warfare as we knew it had just come to a screeching halt.

They couldn't have been more wrong.

A quick look around the world today—more than half a century later—tells us that, but for technology, war has changed little since those hot August days in 1945, and the guerrilla has somehow managed to remain firmly entrenched in seemingly countless battlefields around the globe, with Australia and Antarctica being the only two continents that do not have guerrilla wars going on. As has been the case since the first government was established thousands of years ago, tyrants and tyrannical governments are as commonplace as invading armies, and citizens are still growing tired of being spat upon by those who see them as nothing more than exploitable assets who are powerless to do anything about their plight.

Enter the guerrilla.

A guerrilla war is a war of the people, and, as such, it is one that can be waged by the proverbial average citizen who has had enough abuse and has decided to do something about it from the shadows of the forest and skyscrapers. Armed with rifles, pistols, shotguns, and machine guns—all of which can be garnered by raids on government outposts—and, hopefully, a cache of explosives, the people come together to form a guerrilla band intent on freeing themselves of the chains of oppression or driving off a hoard of murderous invaders.

In the darkness of night, four men ambush a military vehicle by making it appear that their car has broken down in the middle of the road and they need assistance. The truck, they know, has just left an ammunition depot and is filled with machine guns and machine gun ammunition, plus several crates of antitank mines.

A guerrilla war has begun.

This scene is being played out almost daily somewhere in the world, but only a small, select percentage of guerrilla movements ever attain their goal. The reasons are many, but each always comes back to mistakes made and actions not taken by the guerrillas because they simply didn't understand the myriad intricacies of a guerrilla war and how to wage one successfully. The Chechens, minutemen, Afghan mujihadeen, Vietcong (VC), Sandinistas, Red Chinese, and Israelis all understood the formula needed for the conduct of a fruitful guerrilla war, but far more groups did not and fell by the wayside with bullets in their heads. That is the reason for this book.

In my many years of service as a Marine, I sometimes found myself either helping to train a guerrilla force or trying to crush one. Today, I am retired from the Corps but still see guerrilla forces around the world fighting for freedom against corrupt governments or invading armies. And in today's helter-skelter world society, there is simply no telling where the next guerrilla war is going to break out. Tomorrow might see your country—the country you love—invaded by a powerful and deadly army; you will have to fight them in the streets, fields, forests, and alleys. Or perhaps your rights will be stripped from you finally by a criminal politician and his henchmen, and you will suddenly find yourself in pitched battles against hired guns masquerading as military men. Whatever the case, you are going to have to know the ins and outs of a guerrilla war if you expect to be victorious. This book will help you.

Here you will learn everything you need to know in a book that sets down principles based upon thousands of years of guerrilla warfare and demonstrates how those principles are made even more useful by abiding by Mao's Three Rules and Eight Remarks for dealing with the civilian populace, which follow:

RULES

- All actions are subject to command.
- Do not steal from the people.
- Be neither selfish nor unjust.

REMARKS

- Replace the door when you leave the house. (Doors were removed and used as beds on hot summer nights in China.)
- Roll up the bedding on which you have slept.
- Be courteous.
- Be honest in your transactions.
- Return what you borrow.
- Replace what you break.
- Do not bathe in the presence of women.
- Do not without authority search the pocketbooks of those you arrest.

In the coming chapters you will learn how to do the following:

- use mines and booby traps
- plan, set, and execute an ambush
- develop and use leadership principles
- use fieldcraft to increase your combat power
- demonstrate sound leadership traits
- fight in an urban environment

- utilize the principles of *dau tranh*
- select the best explosive for the job
- build and use tunnels and underground bunker systems
- handle and exploit prisoners
- operate at night with great success

Further, you will assimilate countless other skills to conduct a winning guerrilla war.

You will also learn, through detailed accounts, how guerrillas of the past have become victorious, as well as how to avoid the mistakes made by those who have gone before you.

Now you are about to become one of the *Guerrillas in the Mist.*

CHAPTER 1

The Home-Grown Guerrilla

"Stand and face the hounds of hell."

—Vincent Price in Michael Jackson's "Thriller"

The sounds of the humid jungle night filled the little man's ears and head as he made his way silently through the understory, his bare feet feeling their way across the leaf litter as quietly as a centipede's. He knew this jungle well—knew each tree, each shrub, each bird, each bat, each lizard; the jungle was his home. With his dark skin and curly black hair, he was able to blend into his surroundings, and his fluid, easy movements made him seem more like just another animal under the canopy than a man. But this animal was more cunning than many of the others, and he moved with a sense of purpose and urgency, and revenge, a concept none of the other jungle inhabitants knew. And this small, black man had a name given to him by his parents. Vouza. Jacob Vouza.

Stopping beside a tree cloaked in vines of various widths, Vouza grasped them with gnarled hands and scampered up the tree much like a monkey, quickly reaching the lower edge of the canopy and disappearing into the leafy branches. A moment later, had you been a bird, you could have seen his shining eyes peering from the crown of the tree above the canopy and down onto the large stand of palms below, just into the hinterland that lay beyond the softly rushing surf. The clouds scurrying overhead parted then to reveal a brilliant full moon hanging in the tropical night sky, bathing the palm grove in lunar light. Vouza's eyes took in the scene, and his quick mind etched the information into his brain forever—the number of men, how their weapons were emplaced, the layout of their perimeter, vehicle movement, everything. Then he slithered back down the tree and disappeared.

The little man scurried quietly back through the jungle in the general direction from which he had come but avoided tracing his steps exactly; he knew that the men he was spying on may have seen him pass the first time, so he avoided the same route. More than an hour passed before he reached his destination, a tiny, hidden cave with an opening barely wide enough for a thin man to crawl through, an opening concealed by a dense patch of vegetation. Slipping through the brush, he noiselessly entered the cave and was met just inside the entrance by a set of eyes in the darkness—eyes belonging to another man holding a pistol.

"Martin," Vouza said softly in broken English tainted with the dialect of the islands, "I find them." A slight smile appeared below the second set of eyes.

"You always do, Jacob, my friend. You always do," Martin Clemens replied as he nodded in the blackness of the damp cave, his accent thick with down under. "And what have you got for your old Aussie friend tonight, mate?"

The Guadalcanal native then proceeded to describe in detail what he had seen on his mission. Clemens recorded the information in his head for relay to the Allied Forces headquartered in Port Moresby, New Guinea, via the secret radio a commander by the name of Eric Feldt had given him three years earlier in 1939 for just such an occasion. The critical intelligence received by the Allies helped form the plans leading to the U.S. Marine invasion of the island of Guadalcanal in the Solomon Islands in August of 1942.

A week after the leathernecks stormed ashore to commence one of the bloodiest fights in their 167-year history, Martin Clemens, accompanied by his trusty native scout, reported to the Marine commander on the island, Maj. Gen. Alexander Vandegrift. Vandegrift cited both men for heroism and learned that Vouza had once been captured by the hated Japanese, tortured, bayoneted, and left for dead. But the hardy islander was anything but dead, and he soon crawled off into the jungle and eventually recuperated under the care of Clemens. His continued guerrilla activities with Clemens were his way of paying the Japs back for the countless atrocities they had committed upon his people. The Marines awarded Vouza the Silver Star. For decades to come, the Marines continued to pay their respects at every opportunity to Sergeant Major Vouza, a retired member of the Solomon Islands Constabulary. The Marines knew one tough guerrilla when they saw one. He died peacefully many years later, the news of his passing causing Marines around the globe to pause, bow their heads, and remember.

THE HATRED SYNDROME

Man is one of the few creatures on this particular planet that is known to have emotions, with other advanced simians such as chimpanzees and gorillas being the most noted for showing this trait (which is well-documented thanks to the works of such anthropological luminaries as Dr. Jane Goodall), and it is precisely this trait that the guerrilla has relied on for well over the past two millennia to stifle and confound innumerable foes (starting with, insofar as recorded history tells us, Darius's 512 B.C. invasion of current-day Romania, where then dwelled a nomadic, savage people known as the Scythians who employed guerrilla tactics to thwart Darius's attempts to subdue them). One of the strongest and most causative of emotions—a deep-seated, festering hatred of whomever the would-be guerrilla sees as an invader or unworthy and repugnant political entity—is often the single most powerful catalyst to his becoming a guerrilla and his being willing to carry the fight for as long as it takes to realize victory over his oppressor. This is what Sgt. Maj. Jacob Vouza felt in his heart for the savage Japanese invaders who tortured and tormented his normally peaceful people, and his rage allowed him to perform superhuman feats such as surviving what should have been mortal bayonet thrusts to the abdomen. And it is this emotion that still makes the guerrilla, when led by competent and devoted leaders, willing to do almost anything to win, posing a serious threat to those he opposes.

AN ANCIENT ART

But hatred isn't the only factor involved in laying the groundwork for a successful guerrilla campaign. A practical knowledge of history—not so much the who did what when, but more the who did what how and why (the guerrillas' tactical or operational estimate of the situation)—is also demanded of the guerrilla who is bent on winning at all cost. Virtually every successful guerrilla leader of any import in the past two centuries has been a student of the history of guerrilla warfare, with names like Ho Chi Minh, Mao Zedong, Francis Marion, Fidel Castro, Daniel Ortega, T.E. Lawrence, and Vo Nguyen Giap representing only a tiny handful of those who have come to see victory through the eyes of the guerrilla warrior. (Don't concern yourself with these examples being entirely Communist but for Marion and Lawrence; the wise guerrilla doesn't allow the distasteful political orientation of key figures in guerrilla warfare history to taint his devotion to his cause.) Given this, it is important that we examine the military art demonstrated by selected triumphant guerrillas. Once you know who did what, and why, to win the day, you will be able to see the battle before it happens. And every battle you fight as a guerrilla will have been fought before, albeit in some other place at some other time, but the tactical situation will be remarkably similar to some earlier contest between two sworn enemies.

Tyrolea, 1809:
Two Innkeepers, a Priest, and a Peasant

A sea of colorful alpine flowers danced in the mountainside meadows, appealing to the eyes of innkeeper Peter Kemnater and those of his small guerrilla force that lay hidden in the dark woods lining the road south of Innsbruck. The splendor of the snowcapped Alps added to the magnificence of the panorama, but Gen. Karl Philipp Von Wrede, leading a corps of Franco-Bavarian troops with the mission of destroying the guerrilla force that had recently ambushed a team of Bavarian engineers, paid the picturesque setting no mind. This was a mistake, for had he been more attentive to his surroundings he may have detected the small band of concealed peasants waiting in the forest. After a brief but deadly engagement, Von Wrede and his men broke off the advance, abandoned their guns, and retreated. This was Kemnater's first important victory.

However, Von Wrede was not one to give up right away, and he soon led a second force back into the mountains along the road running from Innsbruck to Brixen. Unfortunately, Von Wrede considered the first Tyrolean victory a fluke, and so he neglected to examine the tactics and strategy of the guerrillas, thus leading him straight into a brilliant ambush once again planned and led by Kemnater, this time in a narrow pass that created the perfect killing field. As Von Wrede's army marched through the pass, Kemnater gave the signal to attack. Huge boulders, carefully prepared for immediate use long before Von Wrede's force arrived in the pass, were rolled down toward the helpless soldiers. In the ensuing panic, the mountaineer marksmen opened fire on the disarrayed enemy from their unassailable positions in the trees and rocks above and slaughtered them.

As Kemnater was laying waste to Von Wrede, a second innkeeper, who was the father of the resistance movement, Andreas Hofer, was busy fighting and defeating the invaders around Passeyr (with intelligence provided by a Capuchin priest by the name of Joachim Haspinger), while an otherwise unremarkable peasant by the name of Joseph Speckbacker sacked Innsbruck. These three attacks were well-coordinated by the guerrillas, such that one of the most astonishing victories (from a potential combat power standpoint) ever achieved by a guerrilla force was pulled off. The result was the capture of two generals (and their entire staffs), more than 6,000 infantrymen, 1,000 cavalrymen (along with 800 of their horses), and several cannons.

What mistake did Von Wrede make? And what guerrilla warfare concepts and principles did the Tyroleans successfully demonstrate a keen understanding of?

First, Von Wrede's mistakes.

- Von Wrede had fallen victim to the same fate of many generals of his time, that being a belief in his personal superiority to such a degree that it was manifested as arrogance—he considered himself too smart and too educated to be truly defeated by a ragtag band of unruly peasants. This led to his inability to grasp the importance of understanding how the peasants were thinking tactically.
- Von Wrede believed in the invincibility of his force from a numerical and technological standpoint, and he failed to appreciate the terrain available for use by the guerrillas (and apparently had no understanding of how he could also use that terrain).
- He used no effective advanced reconnaissance forces to tell him of the potentially dangerous pass ahead, and he never believed it was possible for him to be ambushed a second time.
- His lack of a viable intelligence network added to his susceptibility to attack and meant he had no way of knowing that the guerrillas were organized to the degree that they could simultaneously attack three significant targets and do so with great prowess.

Now Kemnater's fortes.

- Kemnater had an acute understanding of his enemy, which, historically, is an advantage common to all fighters, be they guerrillas or regular forces. He demonstrated this by correctly predicting Von Wrede would travel the Innsbruck to Brixen road and not avoid the pass where the ambush was set.

- Lacking any formal military training in the tactics of the day as espoused by most regular European forces—he wasn't burdened by such static concepts as those championed in Humphrey Bland's *Treatise on Military Discipline*—Kemnater operated tactically in much the same way he hunted the clever chamois and red stag; that is, he used the natural features of the land to dictate his movements. Trees, rocks, and the terrain itself all provided him and his guerrillas with excellent ambush points.
- Kemnater, although the beneficiary of little formal education, understood the importance of gathering and disseminating intelligence, and he knew that the enemy was not prepared for or even expecting simultaneous attacks in different locations aimed at critical objectives.
- Finally, he understood that all this meant a much smaller force could surprise a larger, better-armed force and soundly defeat it. He believed.

America, 1755: The Swiss Solution

When Gen. Edward Braddock, a classic British infantry officer with much experience in the European theater, arrived in the Colonies in the mid-1750s, he had no way of knowing that his traditional training and warfighting ideology were to doom him to a stunning defeat in the Monongahela Valley at the hands of a combined force of guerrilla (French) settlers and Indians, and even cost him his life, this despite his numerical superiority, experience, and substantial weaponry. Braddock, unable to comprehend the use of guerrilla tactics, allowed himself and his force of 1,400 to be pinned against the banks of the Monongahela River at its confluence with Turtle Creek by a force of 500 fewer men and severely thrashed. This momentous engagement caused the British hierarchy to rethink its strategy and tactics.

Enter Col. Henri Bouquet, a hired gun of Swiss heritage brought on by the Crown to deal with the increasingly audacious French and Indians. A savvy student of military history, Bouquet did not loathe the French and Indians and their tactics, but admired them to a degree that caused him to abandon the standard tactics he used with such efficiency on the manifold battlefields of Europe in favor of those used with such implacable efficacy by his new enemy. Bouquet would himself become a guerrilla. (It is interesting to note here that the Swiss have a long history of being excellent guerrilla fighters, and to this day every Swiss citizen residing in Switzerland is required by law to maintain a weapon and be part of a militia—unlike the embarrassingly ill-led and trained "militias" that have appeared on the American scene of late—designed to fight invaders with guerrilla tactics. Switzerland is a country rife with a warren of underground bunkers dug into the extreme terrain of the Swiss Alps that will support a guerrilla war no invading army would want to tangle with. The Swiss easily compare favorably with Hannibal's forces, the Vietminh and Vietcong (VC), and the Jews' Irgun, although the latter did adopt a policy of terrorism, which is inadvisable and correctly viewed as morally corrupt by modern Western thought.

Bouquet, like successful guerrilla leaders before him, those he currently faced, and those he would face in the future, saw that the following three basic precepts were called for when fighting a guerrilla war:

- A counterguerrilla force (such as his light infantry regiment) must be armed and equipped with weapons and gear that suit rapid movement in all phases of battle.
- It must always remain dispersed in order to avoid mass casualties.
- It must be physically prepared and mentally willing to pursue the enemy constantly and rapidly to prevent them from counterattacking.

These axioms laid the foundation for his successful operations against the French and Indians, so now let us examine where

Braddock went wrong and Bouquet succeeded (and why other successful counterguerrilla forces, such as Rogers's Rangers during this same period, whose guerrilla warfare principles are taught to this day at the United States Army's Ranger School and to Force Recon Marines, proved effective). This comparison of tactical standard operating procedures (SOPs) can be likened to a comparison of traditional American infantry tactics against the Vietcong and those utilized by American forces willing and able to conduct business just as the VC did. History tells us that the latter—Special Forces, Rangers, SEALs, and assorted Marine units—were much more effective in dealing with the enemy than those grunts who didn't adapt to the situation.

Braddock, probably through his quiescent tactical training via the Continental school of thought, was unable to adapt to the merciless hit-and-run tactics employed by the French and Indians. This made him extremely predictable—his maneuvers easily anticipated and exploited by a crafty enemy who knew when to stand and fight and when to break contact and run. (Although earlier settlers had found ways to downsize units and lessen individual loads to maximize ease of movement, Braddock failed to grasp the tactical importance of this.)

Braddock's lack of knowledgeable scouts to warn of French and Indian forces in the vicinity caused him to either advance to contact—not often the most sound of offensive actions—or be ambushed. He was never able to gain the upper hand with a deliberate attack and simple offensive maneuvers like single and double envelopments.

Bouquet, on the other hand, first and foremost respected his enemy rather than fostering a belligerent disdain for them and their tactics. His first order of business was to study them and learn how the French and Indians engaged British forces, where they chose to do so, and when. This dispassionate understanding of the enemy led to his second strength: because he knew his enemy and how they thought when it came to the tactical decision-making process, he was able to mirror them in mobility, maneuver formations, security, reconnaissance, and task organization. The end result was the French and Indians' reduced ability to strike from an ambush and inflict heavy casualties, quickly break contact and disperse, and use a keen economy of management—just the right number of troops for the mission—and thereby lessen the risk of detection. Finally, as the Germans in World War I would do, he adopted a decentralized philosophy of command as the order of the day, which allowed small unit leaders to make immediate tactical decisions on their own without first begging permission from higher-ups, a policy adopted early on by the Continental Marines, later known as the United States Marine Corps, which is still in use today.

Nevertheless, despite Bouquet's clearly superior grasp of the criticality of light infantry tactics interlaced with guerrilla tricks of the trade, the British army never really caught on or accepted this. On the other hand, the colonists, especially those in New England, were quick to take up guerrilla techniques when they broke from the Crown in 1776. Up against the revolutionaries' emphasis on marksmanship, small-unit leadership skills, initiative, cunning, and solid intelligence gathering, the British were doomed to failure from the start. (On the matters of intelligence and reconnaissance, it should be noted that the rebels, as the British referred to them, or minutemen, as the American's referred to them, used human intelligence to great length, such as when the sexton of the Old North Church in Boston, Robert Emerson Newman, an ancestor of the author, hung two lanterns in the steeple to warn silversmith Paul Revere of the British arrival in Boston harbor. This intricate network would prove to be a maddening thorn in the side of the Crown throughout the war. The rebel Newman was captured soon thereafter and severely thrashed.)

REPUBLIC OF SOUTH VIETNAM, 1950–1975: A DARWINIAN DILEMMA

When the first American Military Assistance and Advisory Group (MAAG) arrived in the Republic of Vietnam in 1950 to do what it could for the ill-fated regime and country, the advisors probably had no inkling of the turmoil and hand-wringing to come. The problem would prove to be Darwinian in nature, with the fittest adapting to the changes and the less fit trying in vain to make the day business as usual and suffering a fate that befell those who went before them. The American experience would prove to be much like that of the French in French Indochina, which ended with the disastrous defeat of their garrison (with more than 2,000 dead and well over twice that number wounded) at Dien Bien Phu by the Vietminh in the spring of 1954. The Vietminh were brilliantly commanded by Gen. Vo Nguyen Giap, who ordered artillery to be secretly hauled up into the mountains overlooking the plain on which the French garrison was situated, a stroke of tactical genius that allowed him to shell the hapless French at will.

Similarly, the Americans would by and large fail to adapt to the survival challenges laid before them in an environment that had bested many earlier invaders. This fact is what is perhaps most perplexing and troubling: despite Vietnam's history of warfare, which clearly shows the pitfalls army after army has suffered there, invaders have continued to march into its emerald rice paddies, screaming jungles, and tentacled rivers but never learned.

General Giap, his devoted cadres, and the seemingly invisible hoards of clever guerrilla warriors, the Vietcong, understood their enemy at all levels and thus were able to engage a virtually gigantic, hideously well-armed and experienced foe with the same tactics used by guerrillas since Scythia. Adding to the combat power of the insurgents were the Vietnamese concepts of time—which is quite different from Western perceptions—and acceptable loss. The Vietnamese do not see a decade as being long at all, and a century to them isn't much longer than a decade. This outlook allows them to undertake siege after siege with no thought toward how long it may or may not take; if it takes a thousand years to win, then so be it.

America lost the war not so much on the battlefields—its soldiers won far more important engagements than they lost—but in the White House, Department of Defense (DOD), and Pentagon. Consider the following:

- Presidents Harry Truman, Dwight Eisenhower, John F. Kennedy, Lyndon Johnson, and Richard Nixon each failed to grasp the tenacity of the enemy. This set the stage for defeat as far back as September of 1950 when American advisors started showing up in country to guide a corrupt and often cowardly Army of the Republic of Vietnam (ARVN).
- Gen. William Westmoreland failed miserably to have his conventional forces adapt to guerrilla warfare. He believed that firepower and technology would win the war for America. He was dead wrong.
- Robert McNamara, the secretary of defense during some of the Vietnam War's most vicious fighting, lied to the American public by saying he truly believed the war was winnable. In reality, by his own admission, he was convinced it was a lost cause from the start but continued to advise the president to send more troops into what had become a meat grinder.
- The American public quickly grew sick of television scenes that showed American soldiers being slaughtered in distant rice paddies, year after year, with no return on their investment. As public support waned, so, too, did the soldiers' chances of victory.
- Most officers on the ground did not understand the dangers of guerrilla warfare insofar as its long-term lethality, and when they did they were often not allowed to fight guerrilla style.
- Because of conscription (the draft) and rampant drug abuse in many units, the quality of leadership was frequently abysmal. The best units proved to be those with few drug problems and disciplined noncommissioned officers (NCOs) and officers who saw themselves as professional soldiers and conducted business as such.

The North Vietnamese, on the other hand, had played their hand masterfully. They understood the nature of the war they were fighting and prosecuted it well in most instances.

- Ho Chi Minh and General Giap correctly anticipated that if

America's troops were kept on the battlefield long enough, the American public would demand that they be brought home. Guerrilla tactics suited this belief perfectly and brought the American psyche into the battle.

- The last three wars America had fought (Korea, World War II, and World War I) were all conventional wars. This left the Americans with few leaders who understood the nature of guerrilla warfare.
- The Communists, knowing that a curious American media would soon dig into the corrupt South Vietnamese government and that the ensuing exposé would reduce public support for the war even more (which is exactly what happened), managed to conceal their own corruption and brutality by denying media access to their much more closed society.
- In most instances the North Vietnamese brilliantly exploited propaganda opportunities, such as the traitor actress Jane Fonda manning an antiaircraft gun in Hanoi.

The North Vietnamese won this war as much on Main Street America as they did in the rice paddies and jungles of South Vietnam and in the skies over Hanoi and Haiphong. Today, an American ambassador once again resides in Hanoi and American companies like Nike pay Vietnamese factory workers 20 cents an hour to make $140 sneakers.

Now let us examine what is arguably one of the most fascinating examples of a successful guerrilla campaign ever waged and see how you, as a modern-day guerrilla leader, can learn from a cigar-champing, bold, lucky, and unfortunately communist revolutionary.

His name is Fidel Castro.

The aggressive counterguerrilla soldier on the ground with a machine gun is often more effective than high-tech approaches.

Even armored vehicles like these M113 armored personnel carriers can be defeated by a crafty guerrilla force.

Guerrilla wars can take decades to come to fruition, as the North Vietnamese knew well.

CHAPTER 2

The Jewel of the Caribbean

Cuba, 1953 and 1956–1959

"History will absolve me."

—Fidel Castro after his assault on the Moncada Barracks failed, 1953.

If ever there was a classic example of a guerrilla war won by the insurgents despite a staggering numerical and technological advantage held by a government and its army, Cuba is it.

Fidel Castro, after failing in his first attempt at revolution in 1953 (after which he and his brother Raul were imprisoned for two years by Batista and freed under a general amnesty in 1955), stepped back up to bat in 1956 and very nearly struck out a second time, his tiny guerrilla band never numbering more than 2,000, and that at the end of the war. (A year or so earlier Castro could muster only a few hundred.) Castro's guerrillas were poorly armed and equipped, to say the least. On the other hand, dictator Fulgencio Batista's army eventually numbered 30,000 and was supported by the United States. This put the force-on-force ratio at 15:1 in favor of Batista. But, to the guerrilla, numbers often mean nothing.

The son of a peasant, Fidel was born in Cuba's Oriente Province, where poverty was the norm as a result of the failed social policies of Batista, with much support and encouragement from American mining, fruit, and sugar interests. He was raised a Catholic, receiving his early education from Jesuit priests, and later attended the University of Havana. Castro showed signs early on of his proclivity toward participating and having a hand in leading guerrilla warfare movements. His first attempt at such an insurrection came in Colombia in 1947 when he was only 20 years old, and it failed miserably. The seed of the guerrilla had nevertheless germinated and began to grow rapidly in the fertile soil of Batista's brutal dictatorship.

After rounding up men and logistical support in Mexico, Castro set sail for Cuba aboard a leaky scow called the *Granma* with 82 would-be guerrillas. Disaster struck immediately as the rusting launch approached Cuban shores with Cuban soldiers waiting. Landing in a swamp and losing most of their supplies, the group was ambushed in a cane field and escaped into the Sierra Maestra—the rugged mountains forming Cuba's spine—with about a dozen survivors, including, remarkably, Fidel, his brother Raul, and revolutionary guerrilla Ernesto "Che" Guevara. With 70 or so of his men dead or captured, things didn't look good for Castro, and the Batista government kept the pressure on. However, the struggling guerrilla leader found welcome friends in the most desperate of Cuba's poor—the precaristas—who had no land to legally call their own. He established a security and intelligence network among them that provided the guerrillas with a more secure base from which they could mount minor attacks, and Castro's guerrilla war took off in earnest.

By 1958 Castro's forces had grown to about 300 mostly unarmed—and those who were armed were, at best, poorly outfitted—peasant guerrillas. Not wanting to bite off more than he could chew—he remembered his almost fatal scrape in 1953 and his 1948 involvement in the Bogota riots, which also nearly cost him his life—Castro managed some small victories against government outposts and, in doing so, gained moral support from more and more of the impoverished populace of Oriente Province. Soon he found himself controlling 2,000 square miles of Oriente and began to see increasing support from certain factions within the U.S. government (Castro had yet to declare himself a Communist, and the CIA had yet to realize that his comrade-in-arms, Guevara, was a dyed-in-the-wool Marxist), which was the last straw as far as Batista was concerned. The result was a massive assault on Castro's stronghold that involved virtually every major asset of the Cuban armed forces, including close air support, massed artillery fires, armor, naval gunfire, and nearly 5,000 troops. It appeared that Fidel Castro was about to meet his end.

In about two weeks' time, Batista's 13 maneuvering units had tightened the noose around Castro's position to an area of about four square miles. Castro now displayed a shrewd understanding of guerrilla tactics by slowly falling back and breaking contact with the army each time it advanced, while using scouts to keep an eye on the soldiers and report on their strength, disposition, and composition. He thereby learned the army was growing continuously weaker because of the extremely demanding nature of counterguerrilla warfare fought in rugged mountains with almost constant rains and ever-increasing disease among the troops. Finally, when Castro decided that the army was at its most vulnerable from its clearly tactically and logistically overextended lines, he and brother Raul simultaneously attacked two key positions. Castro went after one of the most exposed and insecure units and wiped it out, killing more than 600 soldiers and capturing a huge stash of weapons and equipment; the survivors broke and ran. Raul assaulted government positions in the north, which yielded dozens of prisoners—all American and Canadian—in the form of civilian employees, sailors, and Marines. But rather than kill them, Raul effected a brilliant propaganda coup by treating them well and releasing them completely unharmed, an act that placated any would-be American enemies.

Once Batista's army was on the run, Castro again demonstrated a sound understanding of offensive fundamentals by sustaining momentum and maintaining contact with the badly mauled government forces, all the while gaining more and more popular support from the people. Batista fell in January 1959.

Why did Batista fail?

- Batista's first and probably most telling error—although it could be argued that his first true mistake was releasing Castro and his brother from imprisonment on the Isle of Pines—was his failure to pursue with vigor and determination Castro's surviving forces after their ill-fated landing in 1956.
- Batista lacked both initiative and resolve and failed to estimate accurately Castro's willingness to live like an animal under unbelievably harsh conditions of depravation and misery in the Sierra Maestra. Had Batista mounted a campaign like that of Operation Summer—his final but doomed attack on Castro's position in May and June of 1958—while Castro's forces numbered less than 20, he probably would have won.
- The dictator did not understand how politically damaging his brutal handling of strikes and protests by peasants would become to him. American support began to wane and quickly evaporated with the number of newspaper reports filed by American and other Western reporters that spoke favorably of Castro's apparently democratic intentions and charismatic leadership style.
- Lastly, the decaying underpinnings of Batista's corrupt regime lent themselves well to popular support of the guerrillas—who promised "land for the landless"—by the peasantry. This was played upon heavily and effectively by Castro.

And how did Castro manage to succeed after such disastrous beginnings?

- Castro's most powerful asset was his devotion to his cause. By being resilient—and lucky—he was able to keep alive his at first ludicrously inept attempts at guerrilla warfare. Like many guerrillas before him, Castro believed he would eventually succeed, and he did.

- A clever assessment of the near inaccessibility of the Sierra Maestra allowed Castro to stay just out of reach of the government forces pursuing him, even when a major offensive operation was directed against him. Terrain, weather, and vegetation kept the guerrillas alive and the army at bay.
- By settling for small victories at first, Castro slowly built up support, without which no guerrilla movement can succeed.
- Castro understood tactics to a surprising degree. His demonstration of this in allowing a vastly superior foe to stretch itself to the breaking point logistically and spiritually—at which time he attacked—cleared the way for his triumphant return to Havana after only two and one-half years of war.
- And Castro used fundamentals such as keeping an eye on the enemy to determine his strength, composition, and disposition and then struck decisive blows when they were at their weakest. He pressed the attack from there and won.

Castro favored simple tools like grenades to conduct his successful guerrilla campaign.

Today, the once-poor Catholic boy remains in clear power and apparently good health, running a nation that continues to suffer from his totalitarian rule. He has survived eight American presidents, all of whom have made it clear that they wanted him gone.

But he is still there.

CHAPTER 3

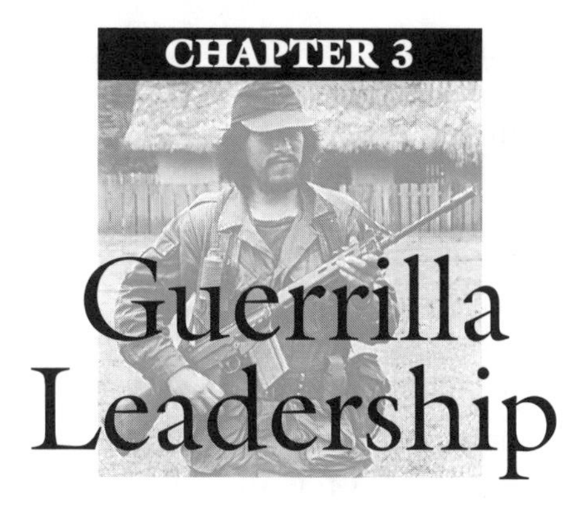

Guerrilla Leadership

"It is the fashion these days to make war, and presumably it will last a while yet."

—Prussia's Frederick the Second
(Frederick the Great) in a letter to Voltaire, 1742

When one examines the history of guerrilla warfare since its inception about five centuries before a woman named Mary, the wife of a poor carpenter, gave birth to a son in Galilee, one finds that of all the intricacies involved in a successful guerrilla war, leadership is key. The best armed, best trained, and most righteous guerrilla army is of little consequence to the government forces it is fighting unless that guerrilla army is led by intelligent, well-trained, daring, clever, and caring leaders who know how to get the very best performance out of the men. History tells us that a guerrilla force with leaders who demonstrate these traits and qualities can defeat even seemingly invincible foes. For instance, when the marauding and undefeated soldiers of the mighty Kublai Khan invaded present-day Vietnam in the 13th century, the Mongols came up against a people with a well-documented history of resisting invaders. In fact, the very first military academy to be founded in Asia was opened in Hanoi earlier in that century, and from it came a tactics book that allowed the Vietnamese to repel the savages who had proven to be, up to that point, unassailable. Leadership was the key to the success of that academy and of the many victories that followed.

GUERRILLA LEADERSHIP STRATEGY

Before we continue, let me say that I am no fan of any form of tyranny, and I loathe what tyrannical governments—especially those based on Communism—stand for and what they are willing to do to their own people in order to achieve their ends. However, as both a student and instructor of military history, tactics, and strategy, I have found that being able to put aside one's personal feelings about this or that form of government in order to examine and exploit the portions thereof that are worth the time and effort makes one a more effective leader. (Wasn't Gen. George Patton keen enough to study Field Marshall Erwin Rommel's insightful treatise *Infanterie Greift An* [Infantry Attacks] to the degree that he was able to defeat Rommel in the North African desert because he had read his thoughts on tactics?)

And as an offbeat example of how some ideas and principles that originate in the enemy camp can prove to have merit, we have America's Social Security program. Enacted in 1935, it makes everyone pay for the welfare of everyone else when they become senior citizens. In other words, the government is forcing individuals to assist in the welfare of the whole, regardless of whether the donators intend to ever use the system themselves. This, of course, is a clearly socialist principle, and socialism forms the underpinnings of Marxism. Nevertheless (although the system is going to have a serious cash flow problem in five years), there are tens of millions of American senior citizens who depend on their Social Security check every month to survive.

Other groups who have employed the ideas, principles, and/or tactics of the enemy to further their own objectives include the Communist People's Army of Vietnam (PAVN) and the infamous terrorist group known as the Irish Republican

Army (IRA), as well as the Haganah, which was a guerrilla/terrorist group of Jews in Palestine during the 1940s when they were fighting for their freedom from the British occupational force in Palestine and for the formation of the state of Israel in its place. (The Haganah supported their rival group's bombing of the King David Hotel in Jerusalem in 1946, which killed 91 and wounded 45, until the support of the Irgun became a political liability and they withdrew it. It is interesting to note that the Irgun was led by a master terrorist who would later lead the entire Jewish nation. His name was Menachem Begin.) While it is true that Vladimir Ilyich Lenin also championed the principles of what we are about to discuss, it will be shown that the system works well when emplaced, employed, and supervised correctly.

Dau Tranh

I will be using Vietnamese terminology to set the stage for this section because the Vietnamese have such a remarkable history of being very successful guerrillas and war fighters in general. For 2,000 years the Vietnamese have continually been invaded and occupied by aggressors from Asia, Europe, and North America, and each and every time they locked into fierce struggle with their foe with the absolute belief that they would eventually win. Results? Yes, they did in fact win every time. They're literally batting a thousand. Any people who can do this over two millennia have my respect and attention, even if their current government stinks.

Dau tranh is the Vietnamese concept of struggle, and it is a struggle based not on Marxist or even Leninist philosophy, but rather a xenophobic outlook linked to the philosophies of both Taoism (which is based primarily on mysticism) and Confucianism, as well as Vietnam's prominent religion, Buddhism. Taoist philosophy stresses the importance of simplicity of action and unassertiveness in attaining goals (it focuses on patience and caution being used to get what you want rather than your attacking the problem without due thought to your options) and was conceived of by a Chinese philosopher named Lao-tzu in the sixth century B.C. Confucianism dwells on the importance of knowing oneself and seeking wisdom through inner reflection and the careful assessment of the opposition (it also came into being in China about the same time Taoism did). Buddhists believe that in all lives there is suffering, but that this suffering can be lessened and eventually almost eliminated through the practitioner's honest and never-ending attempts to become morally and mentally pure. By combining xenophobia (their belief in people other than themselves—all foreigners—as being culturally, spiritually, and morally inferior) with the Taoist, Confucianist, and Buddhist mind-sets, you get a people who truly believe they cannot be conquered and who will do whatever it takes for as long as it takes to eject the invader. These people are the Vietnamese.

But *dau tranh* is much more than a simple struggle. It means a willingness to fight forever against the savage invaders in a glorious and righteous contest in which every person must participate. In Islamic terms it might be compared to a jihad or "holy war." From this concept comes the foundation of the successful guerrilla war: the maintaining of an armed force of guerrillas consisting of those who are fighting for their freedom and rights, led by like individuals who are honored to be and are capable of leading the guerrillas to victory—a force that believes in the sanctity of its goals.

Fortunately—or perhaps not so—you don't have to be a Buddhist, Taoist, or Confucianist in order to utilize the concept of *dau tranh* or its elements, which we will discuss in a moment. *Dau tranh* can be put to good use by any guerrilla force that believes it is right in doing what it is doing and is led by high-quality leaders.

The Key Elements of Dau Tranh

Dau tranh consists of two key elements, both of which must be brought into play if a guerrilla force expects to be successful. The first, *dau tranh vu trang*, is the combat element of *dau tranh*, and it means "armed struggle." It is always a part of the guerrilla war, regardless of who the aggressor is and how that aggressor is conducting his campaign. The second, *dau tranh chinh tri*, is the political end of the concept, but it is a political end that is armed. It is divided into the following three entities collectively known as *van*:

- *Dan van* refers to the civil and administrative activities the guerrillas conduct in areas they have seized and now control. It means "action among the people."
- *Binh van* means "action among the military" and refers to nonmilitary actions taken against the invading army.
- *Dich van* is propaganda. In Vietnamese it translates into "action among the enemy." It is undertaken on both guerrilla turf controlled by the invader and the invader's homeland, if possible.

At this point you probably realize that what we are talking about is total warfare, a concept that, although demonstrated by a regular army, to some degree was practiced by none other than William Tecumseh Sherman during the American Civil War. Because he understood the effects of his waging total war on the populace of the South and its army, and because he exploited those effects to great length, Sherman was able to not only lay waste to the land, but break the spirit of the civilian populace and severely degrade that of the Confederate Army. Of course, there is no indication in Sherman's writings that he was trying to apply *dau tranh* to the American scene in the 1860s, but his actions were clearly along those lines. His grasp of leadership led him to victory just as surely as Grant's grasp of innovative tactics and understanding of maneuver warfare principles—such as his refusal to fall back across the Rappahannock when Lee twice attacked him—led to Lee's surrender at the courthouse at Appomattox.

In contrast, we can look at guerrilla organizations that had the potential, at least in terms of manpower and weapons, to win their respective wars but eventually failed because of a flaw in the foundation of their leader's leadership abilities. Such an example is the Sendero Luminoso or "Shining Path" narcoterrorist group of Peru.

The Shining Path first appeared on the scene in 1980 as the armed branch of the Communist party of Peru. In May of that year, Abimael Guzman Reynoso, a former successful philosophy professor and personnel director at the National University of San Cristobal de Huamanga (and one-time director of the university's teacher training program), declared a "people's war" against the government, with him as the leader of the Shining Path. He liked to be called the "Fourth Sword of Marxism," with his name added to the other three, those being Mao Zedong (Guzman operated with a decidedly Maoist slant), Karl Marx (who coauthored the Communist Manifesto with Friedrich Engels), and Vladimir Ilyich Lenin (the first leader of the Union of Soviet Socialist Republics; he engineered the overthrow and murder of the Romanoffs).

In May of 1980, the Shining Path boasted no more than 200 guerrillas. But the sociopolitical and economic situations within Peru (a murderously corrupt president and 7,600 percent inflation, coupled with a downtrodden peasantry dependent on the coca leaf for survival) were perfect for the guerrillas to exploit, and by 1990 the guerrillas (now turned narcoterrorists) had killed more than 30,000 people. Adding to the Shining Path's ability to wreak havoc was the corruptability of the government forces sent to hunt them and the massive size of Peru, which left them innumerable places in which to hide and operate. Furthermore, the leaders of the future Shining Path were all trained in Maoist ideology in the late 1960s and early to mid-1970s in the People's Republic of China. If ever there was a perfect opportunity for a guerrilla group to succeed, this appeared to be it.

Three factors, however, brought about the downfall of the Shining Path. First was Guzman himself. A remarkably arrogant egomaniac who genuinely perceived himself as being on the same revolutionary plane as Mao, Marx, and Lenin, Guzman considered himself to be the ultimate Maoist guerrilla leader. He was not. Second, Guzman directed the Shining Path to depart from standard guerrilla activities and get into the narcoterrorist trade. This was a strategic error. By doing so, Guzman alienated both the peasants he terrorized in the countryside and the Peruvian middle class and intelligentsia.

Finally, the election of Alberto Fujimori to the Peruvian presidency in 1990 and his reconsolidation of power (rule by decree) in 1992 brought the economy under control. This allowed him to devote more money to building and deploying a better counterinsurgency program. In 1992 Guzman was captured in a raid along with some top Shining Path officials in Lima, a raid that also produced an astonishing intelligence coup in the form of the group's computer files, detailing all their planned actions and who was who within the organization. This allowed the government to rip the guts out of the Shining Path. Since 1990, deaths attributed to the Shining Path have fallen by 85 percent, Guzman remains imprisoned with no apparent hope of ever seeing the light of day again, and Peru's fortunes are improving accordingly.

The Shining Path suffered from a total absence of *dau tranh*.

HARD TO THE CORPS: THE MARINE CORPS CONCEPT OF LEADERSHIP

The prudent guerrilla leader will avail himself of proven leadership traits. To do otherwise will certainly result in failure. Although I am a retired Marine with 20 years of worldwide service in the grunts, reconnaissance, and special operations, and therefore speak with a degree of bias when it comes to whose leadership abilities are best, history tells us that the U.S. Marine Corps' concept of combat leadership is on the cutting edge. This concept is built around selected leadership traits and characteristics that serve the guerrilla leader extremely well. However, without a decentralized philosophy of command—one in which the lowermost commander on the scene is fully authorized and encouraged to make decisions then and there without seeking permission from higher authority, regardless of his rank—these traits, which follow, will be of little use. The guerrilla force operating without a decentralized philosophy of command is akin to putting a governor on the throttle of a Dodge Viper.

Decisiveness

This trait is developed in the guerrilla through a combination of thorough, very demanding training, and the placing of the would-be leader in progressively more challenging positions of absolute authority, including actual combat missions. Take away the challenging positions or the outstanding training, and the guerrilla will fail.

A complete understanding of tactics is absolutely essential for the guerrilla to become decisive. Otherwise, his tactical shortcomings will manifest themselves in defeats on the battlefield, resulting in his men's diminished trust and confidence in him as a leader. To become tactically competent the guerrilla must not only study military history, but the complete history of the people he is fighting, including their culture. The Americans failed in Vietnam, despite a glaring superiority in firepower and technology, largely because the generals running the war lacked a practical understanding of the enemy's concept of *dau tranh*, as well as their concept of time.

Moving the guerrilla progressively up the leadership ladder—starting with comparatively easy missions where success is likely—will build his confidence in himself and his abilities, and his men will develop a like confidence. Basic guerrilla tactics, such as avoiding enemy units with serious combat power and striking those with reduced strength that are vulnerable in hopefully more than one way, are often the best because they are so simple and effective. On the other hand, failed missions must be critiqued thoroughly so that the mistakes made in the battle will be avoided in the future. (A "zero defects" mentality will destroy a guerrilla force before it gets started. Oftentimes, mistakes are excellent learning tools. The guerrilla must, from time to time, be allowed to fail, provided that failure is not catastrophic. The

U.S. Marines, starting during Gen. P.X. Kelley's term as commandant in the early and mid-1980s, began suffering from this problem; it is still one they are wrestling with, but it would appear there might be some light at the end of that long tunnel.)

Dependability

How much you can depend on a guerrilla or a guerrilla unit depends largely on the individual's or unit's character, makeup, leadership, and training. By knowing his men, the guerrilla leader can make an informed decision as to what unit to send on what mission.

Take government armed forces as an example. Responsibilities are delegated based on who is dependable for what type of mission. For instance, underwater demolition work is best conducted by the Navy SEALs, but they wouldn't be the right unit for establishing a guerrilla training base deep in enemy territory; this would fall to the Army Special Forces. And whereas a large-scale airborne assault would best be conducted by the Army's 82nd Airborne Division, the parachute insertion of a small team of men to surreptitiously collect information on enemy strength, composition, and disposition would best be handled by Force Recon Marines.

All this can be directly applied to the guerrilla force as well. By training each guerrilla vigorously and forcing him to max himself out if that's what it takes to accomplish the mission, a guerrilla force greatly increases its combat power. Weakness on the part of one guerrilla can and often does end in disaster. If he can't cut the mustard, lose him.

Loyalty

Although the individual guerrilla's loyalty to the leader is important from the viewpoint of respect, it is more important for the guerrilla to feel loyalty to his unit and for the leader to feel loyalty toward his troops. This way, the unit believes that its existence depends on every man's life and well-being, and the troop feels that it has the respect of its leader. In turn, the troop respects the leader.

The leader must exercise extreme caution to not demand the loyalty of his guerrillas. Loyalty is earned through bravery and a genuine concern for the welfare of every man. Deeds speak much louder than words when it comes to earning loyalty. Once the leader demonstrates to a guerrilla that he is less than interested in that troop's welfare, the guerrilla will never trust or respect the leader again.

Courage

Perhaps the most personal of leadership traits, courage comes from many points. Training in and of itself can't generate courage, nor can tactical superiority, self-confidence, or a technical edge. Courage is quite intangible, but is often born of fear, anger, hatred, and the love of one's brothers.

The leader must prove himself courageous at every turn. He must take every risk his men take and often take it first. Once a leader makes his men suspect or believe that he lacks courage, the guerrilla unit is done for.

But courage must be tempered with common sense and knowledge. Unnecessary bravery often gets guerrillas killed, and few guerrilla units can afford to lose men needlessly. The guerrilla must demonstrate a sense of battlefield intelligence and know-how at all times. Rushing an enemy position with a knife in your teeth is foolish when you could take out the position with a grenade or sniper.

Integrity

This trait is the backbone of the guerrilla leader. Few guerrilla leaders who lacked integrity have ever been ultimately successful.

Integrity is demonstrated when the guerrilla leader does the right thing despite personal cost. In other words, integrity is shown when the leader stands his ground even if he knows he may pay a heavy cost at a later date. It is truthfulness in thought and deed, and it can be shown by guerrilla leaders of any political persuasion. Once his integrity is lost, the leader will never be able to fully recover it, for it is human nature to remember the shortcomings and mistakes of a man longer than his achievements and victories.

Knowledge

This is knowledge not only of tactics and weaponry, but on a broader scale as well. It includes a deep understanding of the enemy's tactics and weaponry, yes, but also encompasses knowledge of the enemy's society, history, culture, government, legal system, and current events.

It is not enough for the guerrilla leader to merely have this knowledge. He must share it at every opportunity with his men and allow them to form their own opinions and ideas as to what happened and why. Classes on every aspect of the enemy must be taught along with classes on the basics of being a guerrilla—tactics, weaponry, and fieldcraft.

Judgment

The guerrilla leader demonstrates judgment in action and everywhere else. This applies to everything from disciplinary measures taken against his own men to the day-to-day handling of personnel matters to decisions made on the battlefield when a mistake in judgment could result in the loss of an entire unit.

Judgment comes from the guerrilla leader's level of maturity, innate leadership ability, experience, common sense, knowledge, personality type, and other less tangible things. It is crucial that the guerrilla leader demonstrate good judgment at all times in all situations, regardless of the stresses applied to him at any given moment. It should also be noted that young NCOs and officers may sometimes demonstrate a lack of good judgment, but they may eventually become better leaders with excellent judgment, provided their seniors apply the principles of solid leadership to them.

Tact

Tact has been described as one's ability to tell another to go to hell in such a way that the man being told begins to anticipate the trip. This is a fairly good definition.

Tact is often the sign of a professional, although there are situations the guerrilla will encounter when tact isn't needed or might even be detrimental. Still, tact shows that the leader need not humiliate a subordinate in front of his peers just to make a point. In most cases, reprimands should be issued in private, whereas praise should be public. The guerrilla troop who feels his dignity has been stripped of him in such a way that his peers no longer respect him or see him as a man of equal worth is a danger to the unit.

Bearing

How the guerrilla leader deports himself is important because he must always appear to be the leader in the eyes of the men. He must lose his bearing only rarely, and then with a plan in mind, which is often to graphically demonstrate the importance of a certain mistake made by the unit. Caution must be taken to ensure that this tactic isn't used too often, which will result in reduced effectiveness.

Screaming, ranting, and "carrying on" as standard operating procedure is the sign of a weak leader who lacks genuine, tangible leadership abilities.

Justice

This trait is sometimes the most difficult to master because it is so fluid. How the guerrilla leader handles problems that require a measure of justice to be dispensed at someone's cost is absolutely crucial to the continued existence of the unit.

There should be two levels of justice in every guerrilla unit—one nonjudicial, the other judicial. Nonjudicial punishment is for minor offenses that did not put another at risk or that did not demonstrate that the offending guerrilla does not care about his fellow guerrillas (such as stealing from another guerrilla). Justice in minor cases should be commensurate with the crime, meaning that you don't execute a guerrilla for some little thing he knowingly did wrong. On the other hand, acts of negligence that put another at risk or that show a serious character flaw that will have some major adverse effect on the unit must be dealt with severely.

In any case, justice must be swift, just, and final.

Initiative

The ultimate guerrilla is an independent thinker operating among independent thinkers. Therefore, every guerrilla leader must strive to take the initiative so that his unit grows to become the most proficient of such units.

No guerrilla should have to be told to do things he already knows he must do. Anyone found to be recalcitrant along these lines needs to be watched carefully and taught to take the initiative. If he never comes around, the appropriate steps must be taken to have him leave the unit. Otherwise, you will have a weak link that may soon become a fatally weak link.

Enthusiasm

In a guerrilla war, enthusiasm can sometimes wane, particularly if the war is a long one. The key to enthusiasm is good leadership.

Every guerrilla must be reminded constantly of what he is fighting for, and he must be rewarded from time to time for his service. The greatest rewards, of course, are personal survival and the movement's victory. This is best accomplished by demanding training that leads to winning small battles, which leads to great battles won with minimal loss.

Wanting to be in the fight is an important trait.

There is more to instilling enthusiasm than merely telling your guerrillas that they should be enthusiastic.

Endurance

The greatest asset a guerrilla leader has when it comes to endurance is a strong personal belief in what he is doing. A sliver of doubt is like a tiny piece of rust. That rust will spread like a cancer and eventually destroy the unit if not handled at the outset.

A guerrilla's endurance can be chipped away from many

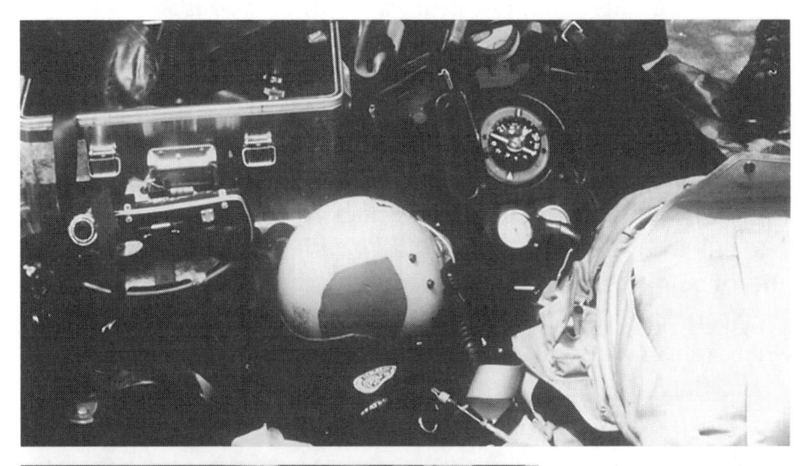

The guerrillas may have to endure a long war against an enemy with advanced gear and tactics.

Small arms are the foundation of a guerrilla army.

angles, and it is the leader's responsibility to make sure those angles are reduced to a bare minimum. Everything from the guerrilla's family's welfare to his own physical and spiritual health must be accounted for. To develop and maintain this trait takes great personal leadership skills, and the leader must show a genuine concern for each and every one of his men and their families if he is to expect his guerrillas to endure the rigors of warfare in a guerrilla unit during a protracted conflict.

Unselfishness

Never must a guerrilla leader be selfish. The insidious character deficiency of selfishness can be one of the most derisive in a guerrilla unit because it tells every guerrilla that you are not in it for them or for the cause but for personal gain and glory. No guerrilla unit has ever been successful that had a selfish leader.

Share everything with your men, including responsibility when appropriate, but never blame them for what is ultimately your failure. You, as their leader, are responsible.

These two scavenging Kuwaiti fighters know the importance of sharing.

CHAPTER 4

The Guerrilla and His Explosives

"Just as one man can beat ten, so a hundred men can beat a thousand, and a thousand can beat ten thousand."

—Miyamoto Musashi
A Book of Five Rings, c. 1643

The modern guerrilla, whether prosecuting a campaign in the concrete jungle of Chicago with former members of the Latin Kings or beneath the triple canopy of the Amazon with Jivaro tribesmen, must have thorough training in the use of every weapon that may become available to him. This includes not only firearms, such as rifles, shotguns, and handguns, and knives, but explosives of all kinds as well. Guerrillas have used explosives for centuries, and with good reason—they work. But there are some intrinsic problems related to their use that must be discussed before we learn how to select and employ assorted explosives in guerrilla warfare.

WHAT WAS THAT BIG BOOM?

The sound of exploding ordnance tends to really get people's attention, and this is the guerrilla's first problem: security in training. Failure to account for and take effective steps to prevent potential enemies from learning of your guerrilla force's activities with explosives is one of the most common mistakes guerrillas make, and it is one that frequently ends with the defeat of the force before it ever gets off the ground. Although two are apparent murderers and terrorists, and one is a convicted mass murderer and the worst terrorist in American history, Ted Kaczynski (the alleged "Unabomber" who was captured by FBI agents in the spring of 1996) and Timothy McVeigh and his partner Terry Nichols, were careless while storing and training with explosives. Kaczynski allegedly kept bomb-making materials in his cabin—a major error—and the Oklahoma City bombers practiced with improvised explosives right on the Nichols family farm—a remarkably stupid idea. Although Kaczynski, McVeigh, and Nichols are not guerrillas but apparent murderers (in McVeigh's case a convicted murderer), the principles here are the same.

Explosives training must be conducted under the strictest security measures. Potential gaps in the guerrillas' security apparatus run the gamut from being caught in a sting operation while buying explosives, detonators, fuses, and caps to being seen or heard detonating the devices at what they thought was a site remote enough to escape detection.

Avoiding the first problem might require the use of improvised explosives by using materials not normally associated with things that go boom in the night. Right now I can spend 30 seconds under your kitchen sink or in your garage or toolshed and come up with all the stuff I need to blow your house flatter than a 3-day-old cadaver's EKG. And if you live on a farm, well, you already know what certain fertilizer and fuel oil mixtures can do. Improvisation is an excellent way of eluding suspicion. Another means of acquiring explosives other than through legal channels is theft; however, explosive materials are always kept in secure locations and are often guarded. The guerrilla's best bet might be to ambush vehicles carrying explosives rather than hitting a static storage site. Beware, however, of explosives that may have taggants or tags inside them. These tags are tiny coded markers that are placed inside the explosive material and can be identified through forensics, giving the other side the ability to tell where the explosive came from and therefore identify a chain of custody that could somehow lead to you.

THE RIGHT EXPLOSIVE FOR THE JOB

The guerrilla seldom has all the explosives he would like available for his use. The truth of the matter is that the guerrilla must often use what is at hand—what he can obtain by hook or by crook. Still, it is important to try to use the right explosives for each job. This economy of force helps to avoid wasting a large explosive charge on a small job that could have been completed with half the charge.

Cyclonite (RDX)

Most frequently used in composite explosives like all the "comps" (Comp A3, Comp B, Comp B4, and Comp C4), RDX is an excellent explosive for the guerrilla because of its great power, versatility, and water resistance. The drawbacks are its sensitivity—it is very sensitive and therefore can detonate accidentally with little provocation under the right circumstances—and its poisonous fumes.

RDX Composites

A3 is mostly made of RDX with a binder/desensitizer made of wax. It is a very good explosive rated below the effectiveness of pure RDX. It is quite water-resistant but, like RDX, gives off poisonous fumes.

Comp B, another good explosive for guerrilla ops, is 60 percent RDX, 39 percent TNT (trinitrotoluene), and 1 percent wax. It is highly water-resistant and offers outstanding power, but it naturally has those dangerous fumes from the RDX.

Comp B4 is very similar in composition to Comp B; however, it contains no wax but rather calcium silicate. It is like an upgraded Comp B and has about the same pros and cons.

Comp C4 is a well known explosive consisting of 91 percent

plasticizer. It is often used in underwater work, is quite moldable, and is very brisant (has great shattering power).

PETN (Pentaerythrite Tetranitrate)

The guerrilla who gets his mitts on this stuff has done good. Extremely powerful like RDX, it is very water-resistant and the fumes it produces are not nearly as dangerous as those of RDX. It's great for underwater bridge and hull work.

Amatol

When you combine equal amounts of TNT and ammonium nitrate you get Amatol. Not a bad explosive, but you have to keep it airtight because of its habit of absorbing moisture like a sponge. Beware of the fumes.

Ammonium Nitrate (AN)

Best used for creating holes in the ground because of its very slow rate of detonation (8,900 feet/second as compared to RDX's 27,400 feet/second), this stuff must also be kept airtight. It is not for use in breaching or cutting.

Military Dynamite

This is often one of the most available explosives the guerrilla can come across. Though not as brisant as RDX, military dynamite is still good for cratering. It is not that useful in wet situations, and the fumes it produces are dangerous.

Commercial Dynamite

Unlike military dynamite, the commercial or standard variety does contain nitroglycerin. Its relative effectiveness depends on its exact makeup and can vary widely from manufacturer to manufacturer. There are four kinds.

Straight Dynamite

Straight dynamite is simply nitroglycerine mixed with an inert filler.

Gelatin

Dissolved nitroglycerin and nitrocotton form the base of this plasticized dynamite.

Ammonia Gelatin

Ammonia gelatin is simply gelatin dynamite with ammonium nitrate thrown in.

Ammonia Dynamite

Ammonia dynamite is nitroglycerin and ammonium nitrate dynamite.

TNT

TNT and military dynamite are not one and the same, as some folks tend to think. TNT has better brisance and is good around water. The fumes are just as bad as military dynamite, however. TNT is the "base" explosive; that is, the explosive all others are compared to when determining their relative effectiveness (RE factor).

Pentolite

When you have a 50/50 mix of PETN and TNT in a single explosive, you have Pentolite. This is another very good explosive, but, as expected, the fumes are bad.

Tetrytol

A combination of 75 percent tetryl and 25 percent TNT

creates Tetrytol. The advantages of this explosive are increased brisance and force along with reduced sensitivity. The fumes are still dangerous.

Nitroglycerin

The extremely sensitive nature of nitroglycerin is not a myth. Although it is one of the most powerful explosives, its uses are limited in its pure form because of this sensitivity.

DEMO GUIDELINES

The guerrilla, when selecting the right explosive for a particular job, must consider the following guidelines if he does in fact have a choice of explosives.

- sufficient power and brisance to accomplish the mission
- resistance to temperature extremes and various storage conditions
- stability in handling—not prone to detonation from jarring and friction
- usefulness in wet climates or underwater
- ease of handling with regard to size, packaging, weight, and so on—not cumbersome
- ability to be handled safely (with certain precautions) by guerrillas (fume avoidance)
- dependability insofar as detonation goes when the initiator fires
- ability to be detonated with a variety of initiators

TYPES OF CHARGES

The charge you select must be the very best available for the job, and the guerrilla leader must strive to attain as wide a variety of explosives as possible so that he has what he needs available when the time comes. Oftentimes, through theft, raids, and ambushes, block demolition charges can be acquired.

Block Demolition Charges

These charges are the guerrilla's mainstay for breaching, cratering, and cutting jobs. They come in cylindrical (roll) and rectangular packages. All are made of either Comp C, TNT, ammonium nitrate, or Tetrytol.

Roll Demolition Charge: M186

Used when you need to wrap a charge around objects with diameters in excess of one foot, the M186 comes in a 50-foot roll and is in fact a rolled sheet explosive. You get half a pound of either RDX or PETN in a foot of the explosive. A complete package contains 15 blasting cap holders (M8). It comes in a satchel. This charge is good for cutting pilons and large trees, but the surface must be free of rust, moisture, ice, and dirt if the adhesive tape on the charge is to hold.

Ammonium Nitrate Roll

The 40-pound ammonium nitrate roll (30 pounds of AN and 10 pounds of TNT as a booster) is an excellent cratering charge and can be used easily in situations that require the charge to be lowered on a cord or rope because of its attached lowering ring. (You can lower it in front of a tunnel entrance/exit from above, into a well, into a storage area from a limb the security forces failed to cut, or what have you.)

TNT Block

Versatile and commonplace, TNT blocks come in 1/4-pound, 1/2-pound, and 1-pound packages. Don't think that a mere quarter pound of TNT is of little use; if you use your imagination and employ such a charge against the right target, it can really impress the enemy. For instance, the guerrillas might stop an enemy food delivery truck en route to an enemy base in order to "collect a toll." While the driver and any passengers are being questioned and shaken down, an unseen guerrilla enters the truck and places a charge inside a large can of coffee or some other easily accessed container, sets the timer, packs some nails around it, then refills the container with whatever he took out. The driver and passengers are then sent on their way. Set to detonate during breakfast when the cooks will be refilling the coffee urns frequently, the device stands a chance of detonating near the enemy as they eat.

Block Demolition Charge: M112

This 1 1/4-pound brick of C4 is versatile and therefore of interest to the guerrilla. It has an adhesive backing that allows the charge to be attached to many types of surfaces, it is packaged in an olive drab wrapper and has excellent brisance, plus it is very good for using on objects with irregular shapes, such as gurters. This charge is most often used for cutting and breaching operations.

Since it is made of C4, it is easily molded for use in special situations. However, this charge will have to be taped to surfaces that either have some sort of residue on them or are frozen or wet. Also, consider the weight of the charge—it isn't a nice, even number. Nice, even numbers make it easy to calculate the required charge size needed to do the job cleanly. Your math is going to have to be up to speed with this charge.

Block Demolition Charge: M118

The name is misleading. This charge isn't a block so much as it is a package of four thin sheets of either PETN or RDX, each sheet weighing half a pound and measuring 1/4" x 3" x 12" This packaging design is for versatility of use, and each sheet can be easily cut to the shape needed and used on curved or other irregular surfaces, from fairly large targets to small targets. Adding to the charge's versatility is the adhesive backing and the charge's ability to be used in underwater demolitions such as hitting subsurface bridge pilings.

Shaped Charges

There are four primary shaped charges in the U.S. inventory, and each is cylindrical with a funneled nose cone on one end. This cone focuses the charge on a small area. These charges are good for creating holes in things such as roads, tarmacs, bridges, concrete, and so on.

M2 Series

The M2A3 and M2A4 shaped charges are both 15-pound charges, but the A3 is made of Comp B with a 50/50 Pentolite booster or all Pentolite, whereas the A4 is Comp B with a Comp A3 booster. The big difference between the two charges is that the M2A4 is substantially less sensitive to things blowing up or firing around it.

The guerrilla can extract the explosive from captured munitions like this Sagger antitank round for later use.

M3 Series

The two charges in the M3 Series are 40-pound charges. They are used just like their smaller sister series.

The M3 itself is made of Comp B with a 50/50 Pentolite booster. The M3A1 is also composed of Comp B, but it has a Comp A3 booster with it.

Again, the Comp A3 makes the charge less susceptible to gunfire and nearby explosions.

These are the demolition basics I'll cover in this book. I won't go into any more detail here because I would end up turning this book into a demolitions book, and I don't want to do that. However, I highly recommend you consult the Paladin Press catalog for books on demolitions and explosives, including improvised explosives.

CHAPTER 5

Guerrilla Warfare Wisdom

Strategy, Operational Art, and Battlefield Tactics

"Strategy, when practiced by Indians, is called treachery."

—An anonymous U.S. Cavalry officer, c. 1865

"Everybody's movin', everybody's groovin' baby . . ."

—The B-52s
"Love Shack"

How is it that certain guerrilla forces are ultimately successful in the face of grim odds (at least in a numerical and technological sense) and others fail miserably? In the end, we see that no single factor determines who is the winner and who is the loser, but rather who knew their enemy best and who used that knowledge to his greatest advantage. These truths are applicable to every military or paramilitary action, regardless of the nature of the combatants.

Of primary importance to the guerrilla is a complete and truthful understanding of how the enemy thinks in three realms: national strategy, operational art, and battlefield tactics.

NATIONAL STRATEGY

Nearly all Western armies first lay out their national strategy (i.e., their goals or policy objectives) and then examine how they will achieve those goals through their national power. National power consists of actions broken down into the following five areas:

- diplomatic
- economic
- technological
- psychological
- military

The successful guerrilla force must first recognize each of these actions and then take the necessary and appropriate steps to counter each one as best it can. All of the most masterful guerrilla forces have understood the importance of engaging the enemy at some level along these five fronts, including American guerrillas during the Revolutionary War, Mao's guerrillas during World War II, and the Vietminh and Vietcong during the French-Indochina and Vietnam Wars, respectively. Yet it is interesting to note here that despite President Kennedy's directed refocusing of American military might on anti- and counterguerrilla warfare in the spring of 1961 and thousands of years of guerrilla warfare history, only a tiny handful of American officers understood the guerrilla. This remarkable lack of comprehension allowed Ho and Giap to defeat not only the French but the Americans as well, even though hindsight tells us that we could have, and should have, anticipated North Vietnam's strategy. Despite the assertions of some authors that the strategy of the North Vietnamese was something entirely new, we now know that the concept of dau tranh was as alive when Kublai Khan invaded what was to become known as Indochina, as it was when France and America followed in his footsteps. In addition, Douglas Pike, a noted authority on Vietnamese warfare, stated that the North Vietnamese guerrilla warfare style is one that "has no known counterstrategy." At first glance, this claim might well appear to be a truism, but the fact of the matter is that America did not have the resolve to win the war no matter what the cost, and the North Vietnamese knew this (a fact that was central to their national strategy). It could have done so with a single hydrogen bomb, but the North Vietnamese correctly anticipated that it was infeasible for the Americans to use such a weapon against so apparently backward a foe; international outrage over the use of such a weapon at such a time against such an enemy would effectively prevent its use. The Communists had preempted four of America's national power elements: economic (sanctions would surely have been levied against America), psychological (America knew it had the ultimate weapon but was prevented from using it by economics and world perception and opinion), technological (the world's most advanced nation couldn't play its trump card), and military (America was prevented from using all of its combat power). This situation can be likened to a vicious fistfight between a smaller man and a bigger man who has a gun, wherein the presence of too many witnesses prevents the bigger man from using his gun.

Let's examine the finer points of what a national strategy comprises.

Diplomatic

The guerrillas must have a clever diplomatic strategy, one that gives the appearance of sincerity but that, in reality, is meant only to frustrate and weaken the enemy over an extended period of time. Look how long America spent negotiating with the North Vietnamese government in Paris before a deal was finally struck—a deal that the Communists, of course, had no intention of honoring, even though the Christmas bombing of 1972 had literally laid waste to North Vietnam. It was part of their national strategy to wear us down at the peace table and finally strike a bargain that would quickly remove all American combat forces from South Vietnam and stop the bombing of North Vietnam. Just over two years after the treaty was signed, North Vietnamese tanks rolled into the American embassy compound in Saigon, virtually unopposed.

Diplomatic strategy works especially well when the guerrillas are ably assisted by ignorant, easily duped, compliant, and extremely naive "journalists" such as *The Washington Post* columnist Richard Cohen. (A better platform for writers of this ilk does not and could not exist.) Cohen's astounding comments on the deaths of the Tupac Amaru terrorists at the hands of courageous Peruvian commandos tell the tale of a man who continues to play directly into the hands of terrorists worldwide. For instance, in a column Cohen authored shortly after the seige ended, he pointedly refused to refer to the terrorists as such, calling them "guerrillas" instead. Guerrillas, of course, engage only military targets; they do not attack residences filled with several hundred innocent civilians and they are recognized by the Geneva Conventions. Cohen willingly lends the terrorists a sense of legitimacy by calling them guerrillas.

Cohen claims that some of the "guerrillas" "may have been summarily executed," but again fails to mention—and even plays down—the Tupac Amaru's long and very gruesome history of bombings, arson, kidnappings, murders, and torture. He goes on to say, through his own special brand of revisionist history and convenient omission of the facts, that the killing of the terrorists may have been "an abuse of human rights . . . more troubling than anything the Tupac Amaru has done." Further, Cohen states that the Tupac Amaru are "hardly a bloodthirsty group." The Tupac Amaru gain much when alleged journalists like Cohen "forget" that some of the group's favorite targets are KFC restaurants filled with children.

Cohen even went so far as to complain that Nestor Cerpa, the merciless and maniacal Tupac Amaru leader who commanded the slaughter of thousands of civilian men, women, and children during his murderous reign, had been shot in the forehead during the operation. Nowhere in the column does Cohen ever mention or grieve for the piles of bodies testifying to the heinous wrath of the Tupac Amaru.

I wonder what Cohen's take is on the 1976 Israeli raid at Entebbe?

It is people like Cohen who see Charles Manson as being a misunderstood humanist, Mu'ammar Qaddafi and Saddam Hussein as being unfairly maligned peace activists, and Abimael Guzman as being a persecuted dreamer.

If a guerrilla group can get someone like Cohen—or better yet, Cohen himself—on its side, it is fortunate indeed.

Economic

Bringing economic hardship on an enemy can be comparatively easy. The key is protraction; that is, keeping the enemy forces involved in the war for a lengthy period of time until it wears on their country's economy to a debilitating degree. The drawback to this is the requirement for the guerrilla leaders to have complete confidence in the guerrillas' tenacity, which must be sufficient to outlast the enemy.

The government of the People's Democratic Republic of Vietnam understood that the American people would never stand for a drawn-out war that cost them tens of thousands of their sons' lives over a period of several years. They knew that the American people were very tired at the end of World War II, which had lasted only four years for America, and were bitter about the result of the Korean War, which ended after a mere three years. In fact, they knew that the longest war the Americans had ever fought (not including the Indian Wars) was their War of Independence, and that was centuries earlier. History told them that time was indeed on their side, and the naive—some would say ignorant, even stupid—politicians and generals running the war only served to bolster the Communists' resolve.

Technological

Seldom will the guerrillas have technology more advanced

than (or even equal to) that of the government forces they are fighting. To effectively deal with this problem, the guerrillas must have a two-pronged approach. First, they must use deception to make the enemy believe that their superior technology is causing tremendous damage to the hapless and frustrated guerrillas and thus continue to devote time, energy, and money to that end, even though in reality it is having little, if any, effect. And second, the guerrillas must seek, find, and exploit gaps in the enemy's technology.

Mobility is often key to the latter. For instance, if a guerrilla force were to purchase or otherwise acquire mobile missile systems such as the SCUD-B and SCUD-C, it could conceivably use the same tactics Saddam's troops used during the Gulf War. Here, permanent SCUD launch sites were left largely undefended and were in fact sacrificial lambs meant to be destroyed by Coalition forces. However, the mobile launchers (on trucks) were maddeningly difficult for the Coalition to locate and destroy because of how easy they were to hide, move, fire, and then hide again. Even daring U.S. Army Special Forces (Green Berets) driving armed dune buggies (Chenowith Fast Attack Vehicles, or FAVs) in the deserts of Iraq had difficulty finding mobile SCUD launchers.

Psychological

Psychological operations conducted by guerrillas are oftentimes among the most effective weapons they can bring to bear on their foe, especially when linked to a protracted conflict.

The doomed Soviet invasion of Afghanistan is a classic example of this. As the war dragged on and more and more Russian boys were sent home in body bags to Mother Russia and their Russian mothers, morale on the front lines (which were extremely vague) and back home in Leningrad, Vladivostok, and Moscow plummeted. The extreme terrain and harsh weather of Afghanistan, the apparent invincibility of the mujihadeen, and the waning support of the Russian populace, who never saw the owning of Afghanistan as being in the Russian national interest, all came together in one big psyop.

And the Russians have yet to learn their lesson. Chechen guerrillas, in their struggle for independence, continually demolished Russian regulars, despite what one would think would be overwhelming technological firepower and logistical ability.

Military

Finally, the guerrillas must be tactically adept and adaptable to the tactics used by the enemy. Also, guerrillas must quickly learn tactics to exploit battlefield conditions such as weather and terrain features. Giap's forces at Khe Sanh during the Tet Offensive of 1968 laid siege to the Marine fire base and were relying upon the monsoons to deny the leathernecks resupply and the ability to mount an effective counterattack. But when the monsoons lifted early, Giap's forces were left very vulnerable to the massive counterattack conducted by the Marines through the use of combined arms concepts (assorted weapons systems being brought to bear on the enemy in such a way that he is put in a dilemma). Giap's bad luck and lack of a plan B resulted in a massive defeat at the hands of some very angry Marines, and as many armies and thugs can attest, angry Marines are bad for one's morale.

OPERATIONAL ART

For the guerrilla, operational art dictates general guidelines for when he is supposed to fight and when he is not. The most rudimentary rule of guerrilla warfare is fight when you have the best chance of winning an important victory and avoid or break contact when you don't. Guerrilla masters have remained true to this axiom for as long as there have been guerrillas.

American minutemen would not initiate an ambush on the British redcoats unless they were quite certain of victory. Japanese soldiers in the Philippines, who remained behind after the war was over (and who were separated from their units), only engaged the enemy when they thought it most to their advantage. The Vietcong almost always lay low until the tactical situation favored them. The terrorists (whom many of the American media like to call guerrillas, not just Richard Cohen) running rampant over much of Lebanon in the 1980s always chose to engage the American and French "peace-keeping" forces in ambushes that took advantage of lethal mistakes made by the Marines and the politicians who controlled them from the safety of their plush Washington offices.

For the enemy commander, operational art is that which links the tactics he employees on the battlefield to his government's national strategy by giving meaning to his operations. You can bet that the government trying to put an end to a guerrilla insurgency is going to do everything in its power to avoid situations where the guerrillas are likely to come out on top and thus move closer to the realization of their goals. Therefore, guerrilla warfare at the operational art level is a game of outwitting the enemy army and exposing decisive, exploitable gaps within the framework of their maneuvers, leadership, communications net, and logistics train.

In order to predict how the enemy's operations will take shape, the guerrilla must both grasp the enemy's national strategy and understand his battlefield tactics. The former is done through the sound collection and interpretation of the enemy's claims released through the media and through various propaganda agencies and mechanisms. The latter is accomplished by closely studying the enemy's military history and dogma, gathering tactical and operational intelligence, and accurately interpreting that intelligence. Once this is accomplished, the guerrilla can formulate his own strategy for engaging and defeating the enemy on all fronts.

BATTLEFIELD TACTICS

This is the nitty-gritty of guerrilla warfare—how the guerrilla locates, closes with, and destroys the enemy in a place of the guerrilla's choosing and at a time he considers to be most advan-

Sometimes guerrillas and their enemy are nearly equal in tactical expertise as well as technology.

Tactics require improvisation. These Marines are using a Japanese soldier's body as protection and a muzzle rest. (Department of Defense photo.)

tageous. Mao understood this as the foundation of guerrilla strategy and stressed the absolute criticality of guerrillas being highly mobile and alert and always at the ready to attack.

Guerrillas conduct combat operations along two primary paths: 1) ambushing enemy missions and convoys and 2) conducting surprise attacks on outposts with the following characteristics:

- They are difficult for the enemy to protect with fire support (mortars, artillery, naval gunfire, and close air) or reinforce with additional manpower.
- They are difficult to support logistically, i.e., resupply is extremely dangerous if not impossible.

Guerrilla infantry tactics must be built around the maxim that a small force can handily defeat a much larger one if the offensive principles of exploiting weaknesses, neutralizing the enemy's ability to react effectively, concentrating combat power, utilizing surprise, and exhibiting boldness are fully developed and employed.

Two examples of how smaller forces defeated much larger and more powerful American forces are the bombing of the Marine Amphibious Unit's (MAU) Battalion Landing Team barracks in Beirut in October 1983 (resulting in the loss of 241 men) and the ambush of a U.S. Army Ranger heliborne insertion in Mogadishu about 10 years later (18 killed).

In the first example, one man (the vehicle's driver) and his support team were able to defeat a huge force in a fatally undefended position by correctly ascertaining that the American government had 1) not learned its lesson in Vietnam regarding overrestrictive rules of engagement (ROE) (the Marines in Beirut were ordered by the White House to not load their weapons and prevent/return fire unless specifically authorized by a commissioned officer—a 22-year-old second lieutenant with 30 days' experience in the Fleet Marine Force could issue orders to return fire, but a sergeant major with 30 years' experience and three wars under his belt could not—a policy recommended by Robert McFarland and supported by the Corps' Commandant, General P.X. Kelley) and 2) had not taken the necessary steps to prevent a repeat of the vehicle-bomb tactics successfully used against the American embassy in that city six months earlier. (And it is interesting to note that dozens of Americans have been killed and hundreds wounded by vehicle bombs detonated at static U.S. armed forces facilities in Saudi Arabia in recent months; will American forces and the politicians controlling them ever learn?)

The second example tells of a unit that failed to adapt to the abilities of what it considered to be no more than an unruly band of hooligans led by an aging warlord (the late Mohammed Farah Aidid) and used fatally (and unnecessarily) brash tactics by attempting a helicopter raid in broad daylight.

This was grotesquely added to by the late and then-Secretary of Defense Les Aspin and two of his primary advisors, Gen. Joseph Hoar (commander in chief of the U.S. Central Command) and the chairman of the joint chiefs of staff, Gen. Colin Powell (who has publicly blamed his subordinates rather than accept any of the responsibility himself). In this last example, Maj. Gen. Tom Montgomery, who at the time was in charge of the Somalia operation, had requested tanks from Hoar and Powell in case their intimidating firepower were needed by the army after the Marines departed. However, after Hoar and Powell failed to sufficiently back Montgomery's request, which resulted in the tanks' being denied by Aspin (who resigned soon after the death of the soldiers in question), Montgomery decided to go ahead with the poorly planned and ineptly led operation without the tanks, a terrible decision compounded by the head Ranger, Gen. William Garrison, who began ordering—by radio; he wasn't actually at the scene of the battle—the on-scene commander (Lt. Col. Daniel McKnight) to do this and that and go here and there in a vicious fight the general wasn't anywhere near. Aidid's guerrillas mangled the Americans with murderous machine gun fire and repeated volleys of rocket-propelled grenade (RPG) fire aimed at the hovering and extremely vulnerable helicopters belonging to the 160th Special Operations Aviation Regiment. Making matters worse was the outrageous lack of contingency planning by the Rangers and Delta Force commandos, who had to wait five hours for relief to come to their aid. And that was provided by Malaysian and Pakistani U.N. forces.

Guerrilla forces that know the enemy commanders won't be held accountable for their failures are made all the more bold by their opponent's deadly displays of cockiness.

Offensive Maneuvers

The guerrilla unit must be able to conduct a variety of offensive maneuvers if it is to be flexible enough to carry on and win a war against a numerically and technologically superior foe. Fortunately, it has been proven time and again that conventional forces, when pitted against a well-led, disciplined, dedicated, and trained guerrilla force, stand a much-reduced chance of ever realizing victory. What better examples of this axiom than the disastrous French and American forays into Indochina and the Soviet Union's fatal invasion of Afghanistan?

But to be victorious, the guerrillas must select and employ the correct tactics for the situation. When considering this, the guerrilla leader must contemplate seven factors affecting his decision. They can be remembered as the acronym METT-TS-L.

- Mission. This is the commander's intent, i.e., what he wants to achieve. It should be simple and clear to every guerrilla, right down to the lowest man in the food chain. Make the objective clear, and then make clear the reason why that objective has been chosen. The guerrilla who understands why he is doing something is more dangerous to the enemy than the guerrilla who just goes through the motions without really understanding the purpose behind the attack.
- Enemy. Here you must inform your guerrillas of everything that is known, likely, or suspected about enemy strength,

composition, and disposition. This includes but is not limited to his weapons, tactics, morale, leadership, logistics, and supporting equipment; his organization and what types of troops he is made up of (infantry, armor, engineers, motor transport, communications, etc.); and what he is up to at the moment (lightly dug in, deeply dug in, on the move along a trail or road, awaiting resupply in a pasture, etc.).

- Terrain. To the guerrilla, terrain means everything. It includes not only the lay of the land but the vegetation, all bodies of water, man-made features, and more. The acronym KOCOA can be used to help remember and plan for the effective use of terrain:

  K = Key Terrain Features
  O = Obstacles
  C = Cover and Concealment
  O = Observation Points and Fields of Fire
  A = Avenues of Approach

- Troops and Fire Support Available. The leader must select the guerrillas for the mission and assign them their tasks. Fire support, such as mortar teams and antiarmor assault teams, must be identified and planned for. Signals for ordering fire support are also brought out.
- Time. Time constraints are important because guerrilla operations are almost always brief and very violent. Ensure that all the guerrillas understand this.
- Space. This is where the leader covers boundaries (control measures) designed to limit advances and flanks. He must perform a careful map study to determine what terrain features act as natural boundaries that can both hem in the enemy and serve to let the guerrilla know that he is near a boundary.
- Logistics. Each guerrilla must have a solid understanding of every logistical concern. Ammunition, medical supplies and facilities, food and water, extraction, and other facets of logistics have to be well thought out and communicated to every man.

Raids

One of the most common guerrilla actions in the realm of offensive maneuver is the raid. In keeping with guerrilla strategy and tactical theory, raids are sudden, unexpected, violent, destructive, and always have a planned withdrawal that immediately follows the end of the mission (whether it was successful or not).

There are any number of reasons why a guerrilla unit conducts a raid. The garnering of weapons, ammunition, communications gear, prisoners, and supplies are just a few. (In many cases, raids are the guerrillas' primary means of resupply.) Meticulous planning, audacity, cunning, and very good intelligence are demanded of the raiding party. Supporting elements must be at the ready to help withdraw the raiders regardless of the situation. Should a guerrilla be separated from his unit, he must possess the skills and will to avoid capture and regain his unit on his own.

Probably the most important part of the raid is intelligence. Guerrillas must never conduct a raid on hopes and wishes, but rather verified intelligence reports that tell of much to be gained by risking a raid. A detailed reconnaissance plan is required, one that produces solid information on the routes of march, assembly areas, line of departure, automatic weapons positions, mine fields, mortar positions, obstacles, individual readiness, reserve forces, command posts, and myriad other factors that will have an effect on the outcome of the mission.

Excellent contingency plans must also be made in case all does not go as planned, and rehearsals conducted with strict adherence to standard operational procedures, under the supervision of seasoned NCOs, must be part of the process.

Frontal Attacks

If reconnaissance patrols report that, without a doubt, the enemy could be easily overrun along a broad front by a sudden, unexpected charge, a frontal attack might be called for. But

These New People's Army terrorists acquired their M16s by raiding Philippine government armories and ambushing government patrols. They are terrorists because of their history of kidnapping and murder.

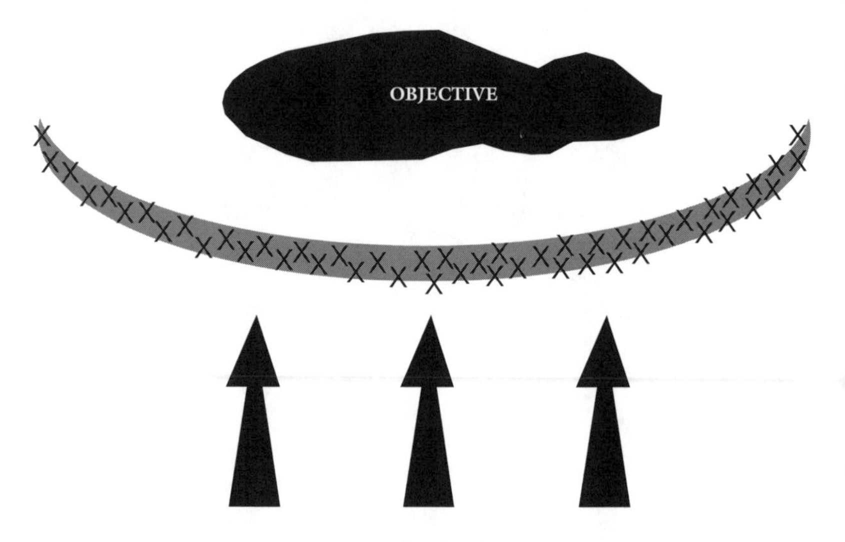

FRONTAL ATTACK

such attacks are always considered a last resort because of the extreme danger involved; if recon reports were wrong and the guerrillas run into a trap, the chances of them being massacred are outstanding.

Speed and surprise are the keys to a successful frontal attack. If the guerrilla commander suspects that the enemy has been tipped off, the attack must be canceled and new plans laid. There is very little room for error in a frontal attack. Even the Japanese, who were often willing to employ their chilling banzai charges at Marine positions during the War in the Pacific, seldom achieved victory because the Marines stood their ground and used interlocking fields of fire with support from indirect fire weapons to mow down the attackers.

Point Penetrations

This was a favorite of the Vietminh during the French-Indochina War and the Vietcong during the Vietnam War. It involves a sudden penetration of the enemy's defenses at a single point, which is quickly followed by the gap being rapidly widened to allow follow-on forces to rush through it in the confusion to strike a decisive blow within the enemy's compound. It is very risky but, as such, often offers great reward when done correctly.

How the rupture of the defenses will be accomplished is situational and will be determined by the leader after a careful reconnaissance and evaluation of those defenses to select the best possible point. The trick is to pick the right place for the rupture (which must take place quickly), instantly widen that gap so that the assault force can enter the compound without being cut down by fields of interlocking automatic weapons fire, and then get to the main target in a hurry and destroy it. Obviously, unless the plan is for the guerrillas to actually gain total control of the objective, they must also have a plan for withdrawal that gets them out of there before fire support and the counterattack force can be deployed.

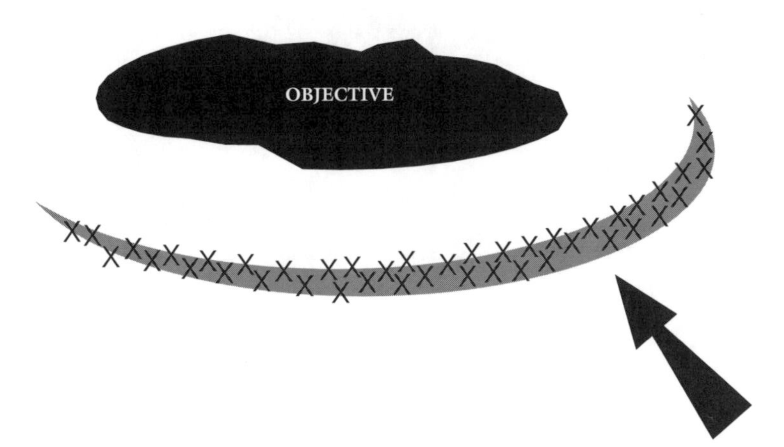

POINT PENETRATION

Single Envelopments

The single envelopment is one of the most useful forms of maneuver for a guerrilla force. Here, a supporting attack by a smaller group of guerrillas engages the enemy at a location along the enemy's defensive perimeter that makes the enemy believe a major attack is under way there. In reality, a larger or more heavily armed force has quietly sneaked into another position along a vulnerable flank that will give them access to a decisive objective inside the enemy's perimeter. When the enemy commits his forces to the supporting attack, the enveloping force strikes quickly and savagely, destroying the objective selected for its criticality.

The following are the keys to a successful employment of a single envelopment:

- a supporting attack that convinces the enemy it is the main attack
- an enveloping force that has avoided detection until it is too late

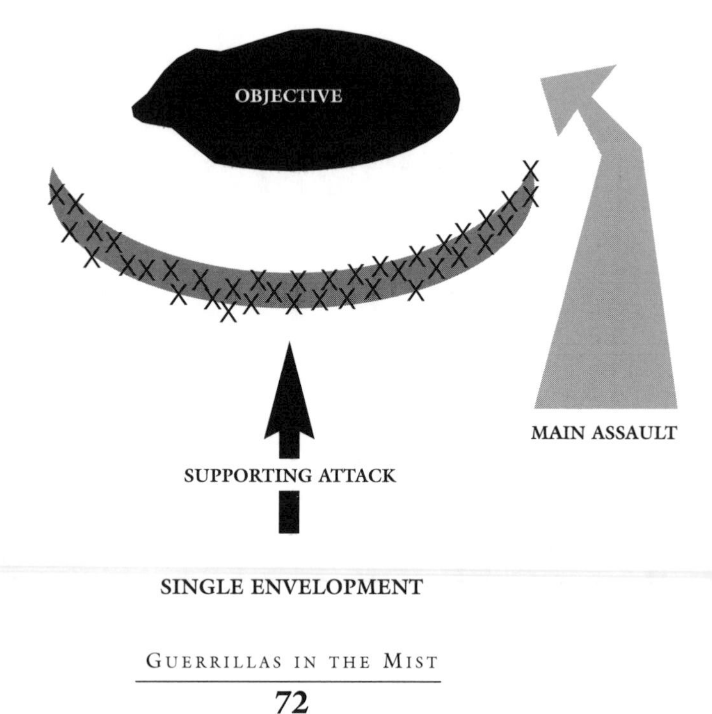

SINGLE ENVELOPMENT

- the selection of an objective that is truly decisive and vulnerable
- an enveloping force that has sufficient combat power to seize or destroy the objective before the enemy's reserve force has the opportunity to counterattack

Should the enveloping force be discovered trying to get into position, the attack must be canceled and the supporting and enveloping forces withdrawn immediately so that the enemy is unable to conduct pursuit or exploitation operations against the guerrillas.

Double Envelopments

Identical to the single envelopment in concept, the execution of a double envelopment simply adds a second enveloping force to the scenario, with the second force attacking another vulnerable and decisive objective within the enemy's perimeter.

The obvious drawbacks to the double envelopment are that the guerrillas must covertly position a second main effort body without detection along the perimeter and that the likelihood of friendly fire increases because of the counterpositioning of the enveloping forces. Also, should one of the enveloping forces get into trouble within the objective, they are going to have to be extracted by either the supporting force or the other enveloping force. Either way is grim. And to leave the force in trouble there without extract would crush the unit's morale by telling each and every guerrilla that he is very expendable and may not be able to depend on help from his fellow guerrillas.

Turning Movements

The insidious turning movement is a maneuver of great value to the crafty guerrilla leader. In this maneuver, an objective important to the enemy that can be made to appear vulnerable and desirable to the enemy is selected for the supporting attack, but that objective is not really the objective at all. This false

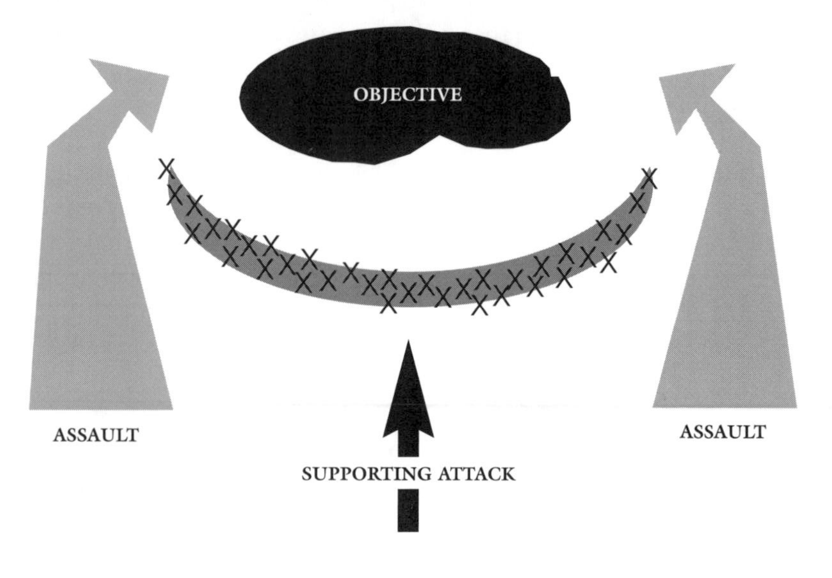

DOUBLE ENVELOPMENT

objective is always within emergency defensive support range of the enemy force occupying the guerrillas' true objective, and it is as deep as possible in the enemy's rear area, or at least well into his main battle area.

The idea is to cause the enemy occupying the guerrillas' true objective to abandon that objective in order to save the false objective from what appears to them to be a main attack to their rear, or, should the enemy refuse to completely abandon it, at least send enough forces to the false objective to weaken the true objective enough to be taken.

These are the basics of tactics, but to fully understand how a guerrilla force must engage the enemy we must learn about mines and booby traps as well as ambushes, all of which will be covered in Chapters 8 and 9. But right now, let's take a look at the master guerrilla himself, a peasant's son from a far away, ancient land.

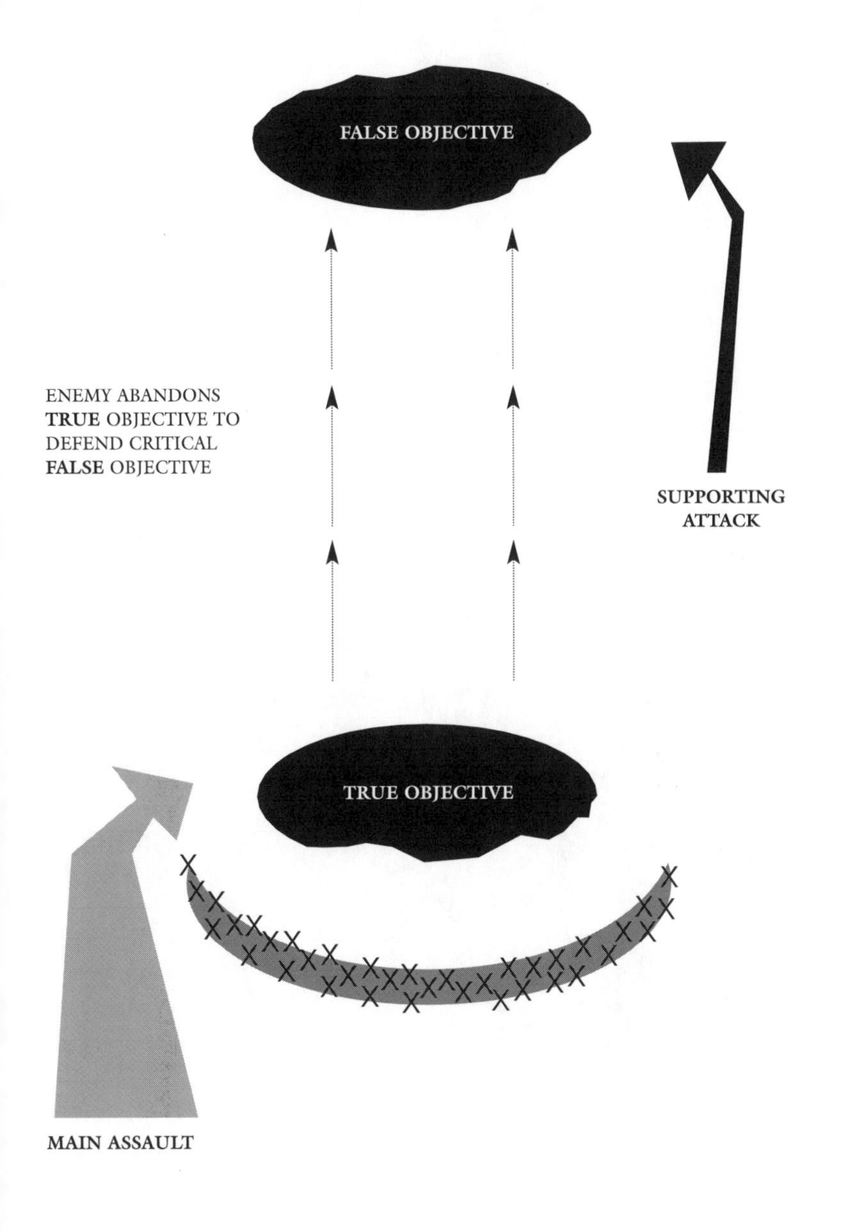

TURNING MOVEMENT

Regardless of the maneuver, fire support may prove to be the critical factor in the end.

Soldier of Fortune's publisher LTC Robert Brown USAR (Ret.) with an Afghan mujihadeen machine gun used, no doubt, in fire support and suppression roles.

CHAPTER 6

The Peasant's Son

"Methods suitable to regular warfare cannot be applied with success to the special situations that confront guerrillas."

—Mao Zedong
"On Guerrilla Warfare"

If ever there was a country ripe for guerrilla warfare, it was feudal China in the 1920s and 30s.

Feudal China? In the 20th century? Indeed, feudal China. Centuries behind much of the Western world when it came to enlightened social thought (in both theory and practice), China at this time was populated with half a billion peasants eking out a brutal existence, made such by rampaging hoards of private armies commanded by savage warlords in cahoots with the land-owning gentry. There was no local government and, therefore, no social services such as those we enjoy today—no police force, no medical services, no running water or sewage systems, no electricity, no schooling except for those who could pay, no nothing. If the marauding mercenaries, merciless winter, or brutal summer didn't kill you, then disease, pestilence, or the culmination of a nightmarish, horrible life did. (It is ironic—and inexplicable to some—that President Clinton continues to extend China "most favored nation trading status," this despite their long and well-documented history—and current policy—of horrific human rights abuses, but he denies the same to Cuba. Clinton clearly understands the axiom "business is business.")

In 1893, in a larger-than-most farmhouse in Hunan Province, a son was born to an enterprising and farsighted farmer (who had managed to secure more land than most of the other peasants, and who in turn was able to afford his progeny an education at a province school that would have been denied lesser peasants) and crafty mother. Historians tell us that this child's interests were diverse, but that his true loves were politics and history.

At the age of 20, his formal studies complete, he was offered an assistant's job in the library of Beijing University, a position that afforded him more than ample time and resources with which to study his beloved politics and history. Given the plight of the Chinese peasantry and his own upbringing, the young man began searching for a means by which he could change China forever. He believed he found the answer in the writings of Marx, Engels, and Lenin, the latter of whom was still alive and in the process of transforming what was recently czarist Russia into the Union of Soviet Socialist Republics. In 1921, he joined the Chinese Communist Party. Five years later, he returned to Hunan Province to lay the early groundwork for the revolution to come through his insistence on sweeping land reforms and the total elimination of the landowning gentry, whom he saw as being the driving force behind the wretched squalor forced upon the peasants.

Mao Zedong had begun his life's work, an endeavor that would change not only the face of China, but that of the entire world.

MAO'S GUERRILLA PHILOSOPHY

Mao Zedong, in his essay "On Guerrilla Warfare," comes clean right away by stating, "Guerrilla operations . . . are the inevitable result of the clash between oppressor and oppressed when the latter reach the limits of their endurance." From this belief, Mao pointed out that a successful guerrilla army must take seven "fundamental steps" in order to achieve its goals. These are as follows:

- arousing and organizing the people
- achieving internal unification politically
- establishing bases
- equipping forces
- recovering national strength
- destroying the enemy's national strength
- regaining lost territories

Arousing and Organizing the People

"If you are planning to hunt a tiger or going to the wars, take some relatives."

—Chinese proverb

Mao did not fashion these seven steps helter-skelter. He gave considerable thought to what order they must be taken in—and in doing so showed an astute understanding of the foundation that must be laid for a guerrilla force to at least have a chance of success. His analogy that the people are the sea in which guerrillas swim is based upon his assertion that the masses must be motivated and organized before they can hope to fashion and employ an effective guerrilla force. Once a burgeoning guerrilla force has the backing of the people it is trying to free from oppression, the oppressor's job becomes many times more difficult. One of the reasons the Shining Path failed in Peru was because the guerrilla-terrorists began a policy of outright terror (linked to the coca trade) directed against not just the elite and the government but the peasants themselves; they alienated themselves among their own people, who in turn began to see the Shining Path—rather than the government they claimed to be fighting against—as the enemy.

Organizing the people is one of the most challenging aspects of guerrilla warfare, for there are innumerable security concerns in doing so; it is often easy for the enemy to entice one of the people to become an informant within the guerrilla force, and an informant in the guerrilla force, if he (or she) takes the appropriate security measures, can be the group's Achilles' heel. For instance, not only was it an informant that led the Peruvian police to Guzman's hideout, but it was an informant who helped Israeli counterterrorist forces to the terrorist known as "The Engineer" on the West Bank, and it was an informant who spilled the beans on an Irish Republican Army terrorist who was gunned down on Gibraltar by the British Special Air Service (SAS).

Recruiting must be carefully done as well. Recruits must be aroused into signing on with the guerrillas, and this can be accomplished by exploiting crimes and mistakes made by the enemy and the government they are supporting, as well as by taking the initiative to care for and protect the locals from harm. Recruits must never be "drafted" or pressed into service with the guerrilla unit, for this will alienate not only the recruit but the entire village from which he came. Besides, few recruits forced to join will ever prove to be highly effective in combat because their hearts probably aren't in the fight.

Villages and towns on the side of the guerrillas must be organized when it comes to active and passive defense against the government. Active defense is risky, yes, because once the village fights against the government forces the might of that government will likely come to bear against them, and this can cause the village to be wiped out or, worse, blame the guerrillas for their pain and suffering. Passive defense is much wiser in most cases because it doesn't infuriate the government if carried off correctly. The idea is to make the government think that the village just wants to be left alone but that it will cooperate if it can. In truth, the village does very little to really help the government.

Achieving Internal Unification Politically

"Let our object be our country, our whole country, and nothing but our country."

—Daniel Webster
June 17, 1825

Politics are to guerrilla warfare as taxes are to the IRS—one cannot exist without the other. Let's face it: the reason you are in a guerrilla war is because you either want to overthrow the existing corrupt government, be it Communist, Socialist, Fascist, or what have you, and establish a democratic government that is truly by and for the people, or because your country has been invaded by a foreign army and you are bent on destroying them at all costs.

The guerrilla army must be united politically and have a single, clear, attainable goal in mind when it initiates hostilities. Guerrillas and potential guerrillas who are waffling or who are unclear as to why a certain political goal is needed must be convinced through education and sound leadership that the political goal of the guerrillas is one of noble purpose, that everyone is going to benefit from the actions taken by the guerrillas, and that it is their duty to work toward that end.

Education is key. The enemy will have a propaganda machine up and running full tilt, a machine designed to lace a few truths with misinformation, half truths, and outright lies meant to confuse, scare, and otherwise weaken the resolve of the individual guerrilla and the civilian populace who are or might be leaning toward assisting the guerrillas. Fighting this propaganda will be a major concern of the guerrillas for as long as the war goes on. Education of the masses and of the individual guerrilla must be undertaken from the very beginning and carried out until victory is achieved, and it must be given on all levels, from one-on-one discussions between a leader and his charge to village and town classes and information dissemination. Caution must be used, however, when dealing with large numbers of people and through mass media. A radio broadcast can easily be traced to the source and attacked with artillery, mortars, aircraft, and a hasty insertion of ground troops into the area, and leaflets written by the guerrillas being found in the hands of civilians will likely quickly result in retaliation by government forces. The protection of the populace is paramount.

The guerrillas must constantly demonstrate to the populace the evil ways of the enemy, and then the civilians must be shown how the guerrillas are capable and worthy of protecting and serving them. To do this, the guerrillas must establish bases that facilitate the conduct of offensive operations.

Establishing Bases

"The conventional army loses if it does not win, the guerrilla wins if he does not lose."

—Henry Kissinger

Establishing bases for a guerrilla army can be a daunting task when pitted against an aggressive and determined counterguerrilla force backed by a government. But it can and must be done. Security and resourcefulness are critical.

Guerrilla bases are covert by nature, never out in the open for the enemy to see with his remotely-piloted vehicles, manned aircraft, and satellites. In fact, the best guerrilla base is the one that does not even appear to exist, with no physical evidence whatsoever being available to the scrutiny and punishment of the enemy. This is the guerrilla base of the American Revolution.

As you will recall from your American history classes, which recounted how many of the Colonists fought the British with a guerrilla army of minutemen who supported the Continental Army, Navy, and Marines (the latter were primarily used as ship's security, boarding parties, and snipers in the rigging), America was born of a guerrilla war. These farmers, merchants, craftsmen, clergymen, trappers, and other everyday folk wore no uniforms and appeared to be simple civilians going about their lives in the towns and countryside, but when the call came they would sneak away to clandestine musters and move to engage the British with hit-and-run tactics that served them extremely well. When a fight was over they would return to their families and jobs and hide their weapons and gear from British search parties. The Vietcong used this same technique nearly 200 years later.

Security is everyone's concern insofar as establishing and maintaining a guerrilla base is concerned. Children are especially vulnerable to being tricked or scared into telling who is a guerrilla and where the weapons caches are, and every precaution must be taken to prevent this. Ignorance is best here—if the children don't know who is a guerrilla and where the goods are, then they can't spill the beans. The guerrillas and guerrilla supporters must strive to operate late at night (when the kids are asleep) as much as possible and use hiding places that aren't easily discovered by children accidentally. It might seem a good idea to forbid the children access to a certain place or area where guerrilla supplies and weapons are hidden, but this can backfire when the enemy comes to town and asks the children where they are forbidden from playing.

Bases established that are more corporeal in nature must still remain concealed or secret, away from the prying eyes of those who would do you harm. They must always be underground or hidden away in what the enemy considers to be some impenetrable region. See Chapter 12 for guidance.

Equipping Forces

"To make war with those who trade with us is like setting a bulldog upon a customer at the shop door."

—Thomas Paine

How to initially equip and then logistically maintain a guerrilla force is often one of the most challenging aspects of being a guerrilla leader. Nevertheless, history tells us that with perseverance and ingenuity the guerrilla leader can come up with the arms, ammunition, explosives, communications gear, food and water, and support equipment necessary to successfully prosecute—at least logistically—a guerrilla campaign.

If America were ever invaded again by a foreign foe—the Japanese were the last to accomplish this with their ill-fated foray into the Aleutians, which were, at the time, a possession of the United States and not yet part of what was to become the state of Alaska—or should guerrilla warfare be taken up against an American government turned tyrannical, we wouldn't have much of an initial equipping problem for guerrillas, unless our weapons, ammunition, explosives, and communications equipment were first confiscated by the enemy. But the Second

These weapons will become the guerrillas' weapons, a principle Mao stuck by.

Amendment is still partially intact despite our socialist politicians' repeated attacks on it through such measures as the so-called "assault weapons" ban (even though "assault weapons" are used in less than 1 percent of all crimes committed with firearms) and turncoat Sarah Brady's personal vendetta against law-abiding handgun owners. The problem would be maintaining those weapons throughout a protracted war and coming up with the spare parts to fix them and ammunition resupply. Food and water would be equally important and might be hard to get from time to time due to enemy efforts at destroying the sources of both—farms and food caches could be put to the torch and water supplies contaminated, the latter of which is a very simple thing to do. Good communications gear (gear capable of operating with a reduced risk of jamming, monitoring, and direction-finding) would probably prove to be troublesome to find and maintain with ease, and support equipment—everything from packs and magazine pouches to mess kits and canteens—will soon wear out and need repairs and replacement.

The guerrilla unit must have the means to repair broken or damaged items and replace those in need of replacement as well as effectively provide the men with all the ammunition they need. Covert supply lines must be established and protected, and this will take great effort (and the North Vietnamese use of the Ho Chi Minh Trail is a good example of finding a way, no matter what the dangers). Local repair facilities must be maintained, too, and these must be underground.

It is likely that raids on vulnerable enemy supply depots will have to be conducted on a regular basis in order to acquire the necessary items, but the guerrillas must be constantly wary of ruses set up by the enemy that are meant to lure the guerrilla into a trap. This is often done by making a supply depot appear ripe for the plucking when in fact it is heavily defended. And the guerrillas must always be suspicious of "targets of opportunity" that appear as manna from Heaven, those being much needed supplies that are left behind or otherwise made readily available by the enemy. Such things are often ambushes in disguise.

Recovering National Strength

"Eternal peace lasts until the next war."

—Russian proverb

This step involves getting back up to speed after repeated hurtful attacks by the enemy. The guerrilla movement that is able to recover continually from endless brutal battles waged over years is likely to be victorious one day. The most powerful weapon the guerrillas have in this realm is a belief that they are right and the enemy is wrong, and that when the sun finally sets, they will be left standing on the battlefield looking over the broken body of their enemy.

But it would be unrealistic to think that the belief in your cause is enough to win the war. It's not. Winning the war will also take shrewd economics, brilliant tactics, the very best leadership, an uncanny ability to exploit mistakes made by the enemy, and much more. And it will take as few mistakes as possible made by the guerrillas. Intelligent decisions and the gift of correct anticipation are requisite at all levels.

One of the wisest things a guerrilla movement can do is make friends with those who might see them as being in the right, and who might be willing to assist them in their struggle. The North Vietnamese Army (NVA) and their guerrilla counterparts in the south would never have been able to continue what the West knows as the Vietnam War without the logistical and philosophical support of the Soviet Union and People's Republic of China. Virtually every weapon system operated by the NVA and VC—except for those firearms captured on the battlefield—was from either the Soviet Union or China. Conversely, the New People's Army in the Philippines was never very successful, partially because they received almost no foreign support, even though they were Communists.

Friendships with benefactors must be cultivated carefully and expertly in order to ensure continual support, and great caution must be exercised in protecting secret alliances. When Admiral John Poindexter and Lt. Col. Oliver North were caught selling weapons to the terrorist state of Iran and using the money to fund the Contras in Central America, they found out how important security is in such operations—the hard way.

Destroying the Enemy's National Strength

"What the hell is going on? I thought we were winning this war."

—Walter Cronkite
An inadvertent broadcast remark
during the Tet Offensive, February 1968

This is best described as taking the battle to the enemy's backyard and hopefully right into his family's and neighbors' living rooms. This isn't done by invading his town but rather his family's, friends', and neighbors' psyche. If the guerrillas can do this, the war is half won.

Up until the winter of 1968 and the infamous Tet Offensive (Tet is the Vietnamese Lunar New Year), most Americans back home held the belief that America was winning the war in

Vietnam. The wealth of lies and disinformation splattered upon the American people by Presidents Johnson and Nixon, Defense Secretary McNamara, General Westmoreland, the joint chiefs and the service secretaries, and many of their minion, all came crashing down upon Joe American and his family when General Giap launched this massive and costly surprise attack on hundreds of targets throughout South Vietnam, from the Mekong Delta up through the Central Highlands to I Corps, Hue City, and the DMZ (Demilitarized Zone). Although it is absolutely true that, despite heavy losses by the Americans, the NVA and Vietcong lost this daring gambit on the battlefield, the Communists won it in America by slapping the American public into the reality of the situation. Suddenly everyone realized that this war we were supposedly winning wasn't even close to being won, and that despite horrendous punishment being meted out by massive American firepower, the Communists still appeared to be as strong as ever and in no way about to throw in the towel. Effectively, America lost the war in 1968. Sadly, it took the government another five years to cut its losses and run. Two years after that, the Communists rolled into Saigon as they always knew they would.

On the battlefield, the best way to destroy the enemy's national strength is to send as many of their sons home in body bags as possible, and to do so on a regular and graphic basis. Through perspicuous but effective tactics and solid leadership, the guerrillas can rack up a huge body count that will demoralize not only the enemy soldiers but the enemy nation as well.

Regaining Lost Territories

"He who has land will have war."

—Italian proverb

It may be important to the guerrilla movement to recapture land lost to the enemy as a matter of honor, strategy, nationalism, and closure. When it is determined that certain territory must be taken back, you must consider that once you take it back

Soldiers like these will often do whatever they think it will take to prevent the guerrillas from regaining lost territory.

you are going to have to hold onto it. Sometimes it is wiser to just deny it or its use to the enemy without actually occupying it with troops. This can be done by making it too risky or by making it less valuable to the enemy.

To make the land too risky the guerrillas must mine, booby trap, and cover the land with indirect fires to such a degree that the enemy makes the decision to simply leave the area alone. To make it less valuable the guerrillas may be able to alter the land in some way that lessens its value, such as when the defoliant Agent Orange was applied to the South Vietnamese jungle so that—at least in theory—the Vietcong and NVA couldn't hide there as easily. Or the guerrillas may be able to draw the enemy away in something like a turning movement because they believe something more important is being threatened by the guerrillas.

In any case, the decision to regain lost territories must be made wisely and shrewdly.

One can learn much from a peasant's son.

CHAPTER 7

Guerrilla Fieldcraft

"Sweet is the smell of a dead enemy."

—Alus Vitellius
at the Battle of Bedriacum, A.D. 69

Fieldcraft includes all those skills the guerrilla uses to make his existence in the forest, jungle, or what have you more efficient, safe, and comfortable. This might include creating a cooking vessel from a section of mature bamboo; finding, reading, and interpreting sign left by the enemy; purifying or filtering water taken from a mud hole; finding a place to sleep that the enemy will not discover; rigging an improvised antenna; building a booby trap; and much more. The guerrilla leader must go to great lengths to ensure that all his guerrillas have developed their fieldcraft prowess to the highest degree; he can never assume that they are proficient at fieldcraft because they are indigenous to the region in which they are operating. Just because a guerrilla was born and raised in a rural environment doesn't necessarily mean he will demonstrate the woods savvy his fellow guerrillas do.

FOOD ACQUISITION, COOKING, AND EATING

Acquiring food is constantly in the mind of the guerrilla leader. There are four primary means of food acquisition, and the use of each will depend on the particular situation of the guerrilla unit in question.

Higher Unit Resupply

If the guerrilla war is being waged on a national level, lower guerrilla units will be able to—at least sometimes—depend on resupply from higher units. If this is the case, they must exercise extreme caution at all times when receiving those supplies. At no time should the supplying unit and the unit being supplied ever come face-to-face; to do so puts both units in jeopardy rather than only one. Unit-to-unit resupply should be done by cache. This is when the supplying unit stages and hides supplies for another unit and then notifies that unit in some way that its supplies are ready to be picked up.

Whatever communication system is used to tell the receiving unit that their food is waiting for them at such and such a location, it must be a secure one. This might consist of simple yet secure radio transmissions, or it might involve some signal on the ground that the receiving unit will be sure to see, such as a common soda can lying in a certain position near a certain rock, or perhaps a discarded vehicle tire being moved slightly to the right of where it usually is. This is sometimes called a dead-drop signal.

Voluntary Civilian Resupply

Civilians sympathetic to the guerrillas will sometimes be willing to supply them with food. Great care must be taken to protect these civilians from suspicion or, should they be found out, reprisals by the enemy.

The guerrillas and their civilian confederates must work out a clandestine system between them so that both parties remain as safe and detached from each other as possible. The village or town helping the guerrillas must never obviously store excess amounts of food in the town (the same goes for a family helping the guerrillas—they must never be caught with suspiciously large amounts of food). To avoid discovery, individuals must transport the food in small amounts to a cache site. When the time is right, a signal is given and the guerrillas will collect the food.

All foods provided by civilians must be paid for, or the civilians have to be compensated in some other way for the risks they are taking. Such mutually agreeable details must be worked out at the local level.

Should the enemy catch a civilian helping the guerrillas and kill that civilian, the guerrillas must pay serious compensation to that person's family for their loss. They must communicate sincere condolences as well, and they must enact improved security measures in order to prevent additional security breaches. The guerrillas should not take obvious, immediate revenge on the enemy unit that killed the civilian; this will only bring more attention upon that family and village. Nevertheless, by waiting a few weeks and then striking the offending unit while it is well away from the village, the guerrillas can, if they feel it might be useful, report to the family and village that the enemy unit that killed the civilian in question has been severely punished. Reassure them that the enemy has no idea they suffered an attack because of who they killed in that village.

Involuntary Civilian Resupply

When the guerrillas are operating in areas they have not yet pacified, it may be necessary to acquire food by involuntary resupply. There are two means of doing this. The first is with compensation, an excuse, and an apology, otherwise known as forced requisition. Here the guerrillas take food from a village's communal stash (never from a single family unless it is very obvious that family has plenty), apologize for the inconvenience, pay fair market value for what they take (or a little more than fair market value, as a show of good will), and then offer an excuse that places the blame for their having to do this on the enemy. This technique must be used as infrequently as possible.

The second way of acquiring food by involuntary resupply is with a ruse attached. This is when the guerrillas steal food from an uncooperative village and do so in such a way that the villagers believe the enemy stole the food. A clever means of pulling this off is to first meet with the village headman or town mayor to "warn" him that you have received reports of enemy troops stealing food from gardens and communal stashes. Advise him to contact you for "assistance" if food starts coming up missing.

A few days or a week later, have your men steal some food and leave tracks like those that would be left by the enemy (have your guerrillas wear boots taken from enemy dead). When the headman reports to you that the enemy has been stealing food, reassure him that you are doing everything you can to prevent additional thefts, but that your unit is small and has a lot of ground to cover and villages to protect. If you capture an enemy soldier, force him to admit to the headman that he was the one who stole the food. Then take him away and promise the headman that the soldier will be punished severely. This technique keeps the enemy in disfavor with the civilians and may very well bring you and your guerrillas into their favor, which is your goal.

EATING

Food preparation and consumption are always a concern to guerrillas, and an important one at that. Any time a guerrilla or guerrilla unit eats, security is automatically lessened because less attention is being paid to security. There are some steps the guerrilla can take, however, that reduce the chances of a security breach when food is being cooked or eaten.

On the Move

It is often advisable for the guerrilla to eat on the move, consuming a little at a time from a pocket filled with something high in complex carbohydrates, simple sugars, and protein. This is especially useful during midday when a large "sit-down" meal should be avoided; such meals in the afternoon can render the guerrilla lethargic, thus reducing his attention to detail and level of alertness. (The sleepy guerrilla who just consumed a big lunch is sometimes said to have fallen into a food coma.)

As is the case with every meal, the meal-on-the-move must be eaten in such a way that no trace of the food is left for the enemy to find; such sign can tell a tracker a great deal. Apple cores and fruit rinds, cellophane wrappers, bones, and all other evidence must be kept with the guerrilla while on the move and disposed of safely and tactically when the situation permits.

At no time should a guerrilla be eating anything while on the move that hampers his ability to employ his weapon, maneuver, or hide. And he should not have to look down at the food he is eating in order to get a hold of it; he should be able to simply feel for it and bring it to his mouth without taking his eye off the zone or away from the direction he is supposed to be watching.

At Brief Halts

While on patrol or otherwise on the move, the guerrilla unit is going to have to stop from time to time, for whatever reason. Stops like these demand that no food be prepared or eaten—the guerrillas are now a stationary target and all eyes and thoughts must be tightly focused on security all the while.

In a Harbor Site

A harbor site is a clandestine hiding spot that a small guerrilla unit (squad-sized or smaller) uses to rest, and only rest. There is no eating or food preparation. Since the harbor site will always be in a spot difficult to reach or detect by the enemy, security can go down to 25 percent if the situation permits. The harbor site demands as little movement within the site as possible.

In a Patrol Base

Food may be prepared and eaten in a patrol base; however, security is always a serious concern. This means that, depending on the situation, up to 50 percent of the unit may be preparing and eating food; the other 50 percent should be tending to security. Cold food (no fires) is safer because of the lack of flames and smoke to alert the enemy of your presence. There are some situations where the guerrilla can get away with tiny fires, but they must be masked from possible detection at all times by being below ground with the smoke being diffused by vegetation above the fire site. Dry hardwood in pieces no thicker than a pencil is called for so that smoke is reduced. The Dakota hole fire lay is a good guerrilla fire lay because the flames are kept below ground at all times. Coals from the Dakota hole can be saved and used in an underground Dutch oven or similar fire lay so that food can be cooking while the guerrillas are out and about.

Avoid allowing individual guerrillas to each have a fire. The more fires built, the greater the risk of compromise. Squad fires are best, and the guerrilla squad leader must be made responsible for the proper use of those fires.

In a Semipermanent Base Camp

The same rules apply in a semipermanent base camp (no guerrillas ever operate out of a permanent base camp; the risk of maintaining such a static site are too great) that apply in a patrol base. Two of the greatest dangers guerrillas face insofar as detection is concerned when operating out of a semipermanent base camp involve the disposing of food waste/containers and the denuding of the surrounding area by guerrillas foraging for fuel for their fires.

Food waste disposal is made less of a concern by efficient preparation and consumption, i.e., cook only what you intend to eat and eat everything you cook that is edible. Bones and other inedible waste must be disposed of surreptitiously, well away from the camp. Large dump sites are out of the question. This leaves disposal in sites such as deep rivers (for waste that will sink) and one-man dumps. A one-man dump is a tiny hole dug by each guerrilla in which he places his food waste. These dumps are always well away from the camp, and the guerrilla takes care to cover the site in order to make it appear that nothing has happened there. Such holes should displace no more dirt and detritus than necessary.

Booby traps can do some good when placed near a dump site. Some counterguerrilla units will use dogs to sniff out dump sites. The more dogs and enemy forces injured or killed by booby traps at dump sites, the better. The mental stress experienced by the dog handler when searching for a dump site is bad for their morale.

The gathering of fuel for evasion fires was one of the problems we always counseled our students on at the Navy SERE (Survival-Evasion-Resistance-Escape) School in Maine. Just a few men could quickly make a very noticeable impact on the flora in a small area when gathering fuel (tinder, kindling, and the more bulky fuel) for a fire. We taught them to never gather the material they need anywhere near where they intended to build their evasion shelter, and that when they did collect twigs, bark, and small branches that there must be no evidence of that fuel having been removed from where it was. These same principles must be applied by guerrillas.

In some situations the guerrilla may be able to locate and procure alternate heat sources for warming or cooking his food. If at all possible, the guerrilla should have in his kit a small single-burner stove that is capable of using a variety of fuels, such as white gas, Coleman fuel, and gasoline. Stoves that can use only one type of fuel are to be avoided if possible because of the obvious logistical restraints they force on the guerrilla. One problem caused by these stoves is maintenance. Preventive maintenance is crucial in order to avoid unnecessary glitches. The stove must be cleaned regularly and thoroughly, and spare parts must be kept on hand.

As another alternative, that old standby, Sterno, is still a viable and often advisable means of heating food. A single can, if used right, can last quite a while and heat many meals.

During the Gulf War, my Marine infantry unit was introduced to a new type of food warmer. Called an MRE (Meal, Ready to Eat) heater, it consists of a thin plastic sleeve a little wider and longer than an MRE package with a soft, flexible wafer of some material in it that, when soaked with a little water that is poured into the sleeve, quickly reacts with impressive heat. Hydrogen gas is produced as a result of this chemical reaction, so caution must be exercised not to have any open flames near the heater when it is in use.

There are still various forms of "heat tabs" available that burn when a lit match is applied to them. A major advantage of having chemical heating mechanisms such as these available is the fact that they can be carried right along with the guerrilla, thus eliminating the need for building a fire.

If logistics permit, each guerrilla should have his own cooking vessels and utensils in his kit; however, this isn't always the case. In many situations, the guerrilla is going to have to fashion and use items provided by nature.

If the guerrilla is fortunate enough to be operating in areas with stands of mature bamboo, the world's thickest and often

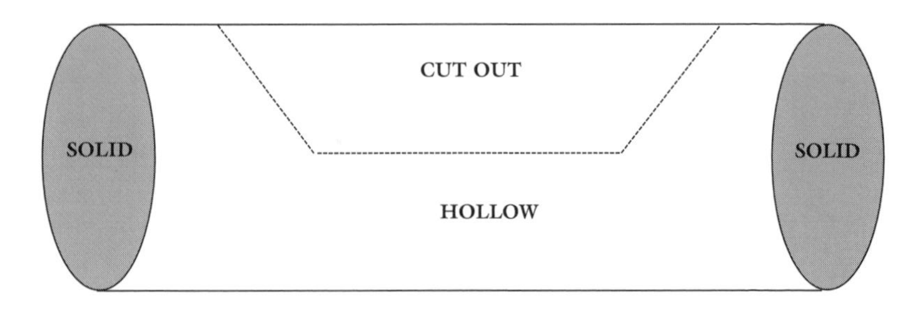

BAMBOO COOKER

most useful grass, he can use a section of the plant to make a vessel for steaming or boiling food. By cutting a rectangular lid in the center of one section, he can stuff food into the cavity and add a little water. He then replaces the lid and sets the bamboo over a fire. As an alternative, he can take a large section of bamboo and cut it off in two places—below the upper joint so the top is open and below the lower joint so the bottom is solid. Then he can set it up on one end between two burning logs or some rocks to cook the food he has placed inside.

Large green leaves like that of taro can be wrapped around food items, which are then cooked over coals. The guerrilla places coals in a shallow hole, places rocks over the coals, sets the leaf-wrapped food on the rocks, and then covers the whole deal with dirt. This is the guerrilla's Dutch oven, a system that allows him to cook food with no further effort while he tends to other things. Also, those same leaves (or green leaves similar to them) can be wrapped around food and shoved directly into the coals of a burning fire.

HOBO STOVE

A medium-sized steel, heavy aluminum, or tin can is easily fashioned into what is called a hobo stove. Such a device can last for many meals if made correctly and cared for.

SLEEPING

Where, when, and how a guerrilla unit sleeps is of critical tactical importance. As a general rule, most sleep will be taken during the day, since most guerrilla operations will be conducted at night. Although caution pervades sleeping arrangements, the guerrillas must get sufficient sleep if they are to be at their sharpest during attacks, patrols, and other activities.

On the Ground

Most of the time, guerrillas will be sleeping on the ground, with one exception being in wet areas like swamps, pocosins, marshes, and so on, which require either the construction of platforms or the use of hammocks strung between trees.

A guerrilla should never have to sleep with nothing between him and the ground except the clothing he has on. A ground mat, poncho, tarp, or similar item should be issued so that a buffer separates the guerrilla from the damp and/or cold ground. Remember that the guerrilla must be able to get some sleep; the guerrilla leader must use whatever plausible means he can come up with to make his fighters more comfortable.

In anything except subtropical or tropical climes, a sleeping bag of some kind is going to be required. In warmer, temperate zones, a sleeping bag can be constructed out of a poncho and a liner, but in areas with snowfall and colder temperatures, a genuine sleeping bag will be needed. As is the case with most gear guerrillas acquire, sleeping bags, tarps, ponchos, liners, and other things will be stolen or captured from the enemy. In friendly areas where the indigenous people favor the guerrillas, cottage industries can be set up to manufacture these items.

A technologically advanced enemy may have thermal imaging systems mounted in aircraft. In this case, guerrillas must sleep in the fetal position at night in small groups; this can make them appear to be a small herd of mammals such as deer. The vast majority of mammals sleep curled up this way to facilitate body heat retention. The guerrillas on guard duty will ensure that no sleeping guerrillas stretch out in the linear.

Above the Ground

When the guerrilla is required to sleep up off the ground, he may have up to three choices, not including sleeping in a vehicle, which is inadvisable from a tactical viewpoint.

First, he can string a hammock between two trees. Most hammocks are fishnet style, which makes them light and compactible. He can string a poncho above the hammock for rain protection (some commercially available "jungle" hammocks have poncho roofs as part of the system). A disadvantage of the fishnet-style hammock is that gear tends to get caught in it. The guerrilla using this kind of hammock should be taught to hang his combat gear on the tree at his head for ease of retrieval should he need it suddenly. Another disadvantage of the hammock comes with cold-weather use. Just as the surface of a bridge freezes before the road surface does because of the cold air passing beneath the bridge, the hammock will have cold air passing underneath it, causing the guerrilla to become colder faster.

The guerrilla can also construct sleeping platforms if time, equipment, and nature permit (enough satisfactory vegetation). Although most types of wood can be used to build a sleeping platform, the easiest to use is bamboo because it is hollow (easily chopped down) yet strong. Training in lashing is required.

The final choice might be sleeping in a tree. There are obvious dangers here, such as falling out of the tree and not being able to escape easily should you be discovered. Still, some trees offer fair protection as sleeping locations. Always use a safety rope to tie the guerrillas into the tree with a quick-release knot. Make sure no sign has been left around the base of the trees being used or in the general vicinity.

WATER FILTRATION AND PURIFICATION

The accounts are many of both guerrilla and conventional forces that fell prey to impure water and lost their struggle because of it. Guerrillas must exercise strict adherence to water purification practices, since so many natural pathogens and man-made contaminants are found in water sources. Clarity is never to be considered a reliable indicator of purity; many contaminants are invisible to the naked eye. Nor is the remoteness of the water source to be considered a reliable indicator; even the most remote streams are likely to contain some natural contaminants, and clouds formed by water vapor originating over contaminated water or containing airborne contaminants (acid rain and other pollutants) can be dangerous.

Filtration

Filtration involves the removal or substantial reduction of suspended particulate matter, such as silt and detritus, from water. Filtration alone, although good for improving the appearance and perhaps the taste of the water, is not enough; it does not remove toxins.

Primitive filtration can be accomplished by suspending a trouser leg from a tree branch with the hem down and a knot tied just above the hem. Fill three-quarters of the leg with alternating layers of rock, sand, and grass (two layers of each). Pour the water to be filtered into the top of the leg so that it drains down through the layers and drips into a container placed on the ground below the knot in the leg.

Two shirts stretched horizontally tight, one above the other, can have water poured through them. Place a container directly beneath them to catch the water as it comes through.

You may be able to simply allow standing water to settle for as little as an hour or so in order for the suspended matter to settle to the bottom of the container.

Purification

Purification is the removal or destruction of harmful pathogens and other materials in the water. Some modern filters available at camping stores and from outdoor equipment suppliers do in fact purify water by removing nearly all known natural and man-made contaminants, including viruses, bacteria, dinoflagellates such as pfisteria, and other debilitating toxins like *Campylobacter jejuni* and *Giardia lamblia*, as well as heavy metals from industrial waste and fecal coliform bacteria, which is common in the water and sediment layers of rivers and streams meandering through agricultural areas.

Boiling water at a rolling boil for a few minutes kills all natural contaminants but may leave synthetic toxins. Still, boiling is better than nothing.

Water can also be purified by chemical means such as bleach or iodine and chlorine tablets. Many conventional armies use these tablets for combat units operating along the forward edge of the battle area. Guerrillas should carefully search bodies and prisoners for these tablets, which they can take and use. Warning: I once operated in a counterguerrilla unit that kept highly toxic tablets that looked just like iodine tablets in a genuine iodine bottle in their uniform pockets in case of capture or death. Make sure you give a prisoner from whom you have just taken what appears to be iodine or other purification tablets a canteen of water containing one of these tablets first. Let him see you place the tablet in his water. If he balks, you know the tablets are bogus.

Water Collection Sites

Water must be collected covertly. Entire guerrilla units must not show up at a single water source to collect water. A few guerrillas should take canteens from the whole unit and fill them from a position of safety. This way, if they are compromised, only a few guerrillas are in trouble rather than everyone.

The guerrillas must be careful not to use the same exact collection point more than once. An enemy unit watching the river (or other source) might easily spot some guerrillas getting water somewhere and allow them to leave without being attacked. Then they can set up an observation post to see if the guerrillas return to that collection point. If so, again the enemy allow them to leave and later mine or booby trap the site. Or the enemy might let them leave and then send in a sniper team to track the guerrillas on their way back to their unit. The rest you can imagine. Another option the enemy has in this instance is to do a thorough map study of the surrounding area in order to deduce where the guerrilla patrol base might be, then send in a recon team, conduct aerial reconnaissance (manned or unmanned), or drop some listening devices in the area. Some counterguerrilla forces might even contaminate the water if they think it might harm the guerrillas.

How the guerrilla intends to use the water he is taking will determine where he will take it from. For instance, drinking water comes from the point farthest upstream. Downstream from there is the point for cooking water, followed by utensil and cooking vessel washing, clothes washing, and, if necessary, vehicle washing (a rare event for guerrillas). Soaps must never be used unless they are the biodegradable kind that environmentally aware backpackers use—such soaps leave no bubbles or other signs of pollution.

NAVIGATION AND MOVEMENT

Guerrilla forces are usually indigenous to the region in which they are fighting. Nevertheless, the guerrilla leader must make every effort to ensure that each guerrilla is not only very familiar with the lay of the land but also knows how to move across that land without being detected. The guerrilla must also be adept at using technical (map and compass) navigation techniques as well as primitive means to get from place to place; the time may come when he must move out of his home area into a region unfamiliar to him, be it as a semipermanent relocation or to link up with another guerrilla force in order to conduct a joint operation.

This book does not contain a detailed explanation on technical and primitive land navigation, since this would require a section the size of a book itself. All the land navigation (and survival) techniques the guerrilla needs are found in *Wilderness*

Wayfinding: How to Survive in the Wilderness as You Travel and in a video, *The Ultimate Outdoorsman: Critical Skills for Traveling, Surviving, and Enjoying Your Time in the Wilderness.* Both are available from Paladin Press.

Elusive movement techniques, however, shall be covered here.

Movement

Mao Zedong, when writing about yu chi chan (guerrilla warfare), said that guerrilla strategy must be based primarily on alertness, mobility, and attack. The first two are crucial to achieving the third.

Anyone, regardless of his background, can learn to move undetected. Whereas it is true that many guerrillas are born out in the countryside and have learned to move quietly through the woods while hunting, a new guerrilla born and raised in a city can also be taught to slink through the forest without a sound. And country boys can be taught to improve their movement skills. (I am a perfect example of this. Although I was raised in the woods of Maine and in the backcountry of old south Florida and the Everglades when this region still had huge tracts of uninhabited wilderness and was quite proficient at sneaking about by the time I was 13 years old, I learned more as a reconnaissance Marine by paying attention to people like Stan Iramk, my point man, who was raised in the Palau jungle in the western Pacific; Pat Halling, a strapping country boy from the sugar beet fields of northern Minnesota who knew how to run a patrol better than most men; and Todd Ohman, my recon team machine gunner.)

The guerrilla must always be watching, listening, and learning. The day the guerrilla feels he has nothing more to learn is the day the guerrilla war is lost.

The Trail Fallacy

I recall being taught in the Marines that you must never use trails because of the increased likelihood of being ambushed.

Although it is so that the more often one uses a trail to get from one point to another the greater his chances are of being ambushed, the truth is that all guerrilla forces and all conventional forces use trails extensively. The reason for this is expediency. It is often necessary to move quickly from one place to the next, and trails afford you that ability. The key is to use trails as infrequently as possible and, when using them, to reduce the amount of sign or impact left on that trail.

To reduce sign (whether on a trail or off), the guerrillas should travel in small groups with the lightest possible loads, and preferably travel during times of darkness and rain. The fewer feet on a trail the fewer footprints left behind. The lighter the individual load, the less the impact of the foot on the ground. Rain helps wash away sign, and darkness allows a certain degree of visual safety.

Small trails are preferred to larger ones because they are more difficult for the enemy to detect. Guerrillas can use game trails extensively in many cases, provided they adhere to the aforementioned rules. It is wise to have a large guerrilla force break up into smaller groups and move independently via small trails and then form up at a predetermined, secure location for the attack when close to the objective. It is worthwhile to train dogs to run point and detect booby traps and enemy ambushes with their sensitive noses.

Interval

The distance between one guerrilla and the guerrillas to his immediate front and rear is called interval. Interval is determined by the tactical situation: you want enough distance to prevent or lessen the chances of two or more men being wounded or killed by a single booby trap or ambush, but you also want to be close enough so that eye contact can be maintained and mutual support is available in case of trouble. Terrain, weather, vegetation, weapons capabilities, and the level of tactical proficiency the unit has all play a role. In any case, the interval should not be so great that one guerrilla cannot effectively communicate with the guerrilla in front of or behind him with hand signals.

Hand Signals

Hand signals should be simple and easily understood. For instance, pointing at your eyes means you see something, and then pointing in a direction means you see something in that direction. Next, flashing numbers with your fingers means that something is so many yards away in that direction. This signal might be followed by another numerical signal indicating how many of those things you see, which is followed by a signal for what those things are. The latter might be forming a handgun with your hand, which would mean enemy troops. In less than three seconds one guerrilla can "tell" another guerrilla that he sees five enemy soldiers 200 yards to the east and never have to open his mouth or risk detection by moving back to the other guerrilla.

Time Together

Guerrilla leaders must do everything in their power to keep small units together as much as possible. The more time a small unit spends living and operating together, the more efficient they will become as a team. After several months of combat together, a small team of guerrillas can seem to read each other's thoughts based on how they are moving, i.e., their body language. I have operated in units with this ability and can assure you that they are highly effective.

Minimize personnel turnover. Everything should be done as a team—sleeping, eating, training, rehearsing, and fighting. If a personality clash cannot be worked out within the team, move one of the guerrillas involved to another team.

Simple Rules of Movement

The following are some simple rules to make movement more tactically sound:

- A guerrilla must never use vegetation to pull himself up a slope; pulling on vegetation leaves additional sign of the guerrilla's passing.

- Stealthy scouts must be used in advance of the main body, and a reliable communications system must link them.
- Before moving out, the guerrilla must silence all gear with camouflaged (dark green, brown, or black) masking tape.
- Trails being avoided by villagers and townfolk are being avoided for a reason—they are unsafe for some reason, i.e., they are mined or booby trapped.
- Night movement must be practiced more than day movement; night movement is more difficult and the majority of guerrilla operations will be conducted at night.
- When using a trail system, use alternate routes to avoid patterning.

Alternate Movement Techniques

Besides moving over land on foot, the guerrillas should attempt to take advantage of rivers and other bodies of water. Watercraft can be an excellent method of moving supplies and infiltrating troops.

High Speed Cast

While operating in the Philippines, one of my reconnaissance unit's favorite insertion methods was by bonka boat—dugout canoes with outriggers and outboard motors operated by indigenous people like the Negritos and the Filipinos themselves. The boats are extremely common and therefore seldom attract attention. By having the boat operator run the boat at high speed (which is just about the only speed they run at) along a shoreline, a team of men can "high speed cast" into the water without the boat's having to slow down. They can then infiltrate from there.

Rung and Snorkel System

Slower craft can also be used, but like any boat they can be searched. In warmer waters the guerrillas can hide beneath a boat that has been specially rigged. Attach rungs on the bottom of the hull, and at each rung drill a narrow hole in the hull. Into this hole is inserted a snorkel with a short extension attached that runs up underneath a seat so that it can't be seen from inside the boat. The guerrillas hang on to the rungs and also attach a safety line between them and the wrung in case they lose their grip. I have used this trick many times and have yet to be caught by a search team investigating the boat midstream.

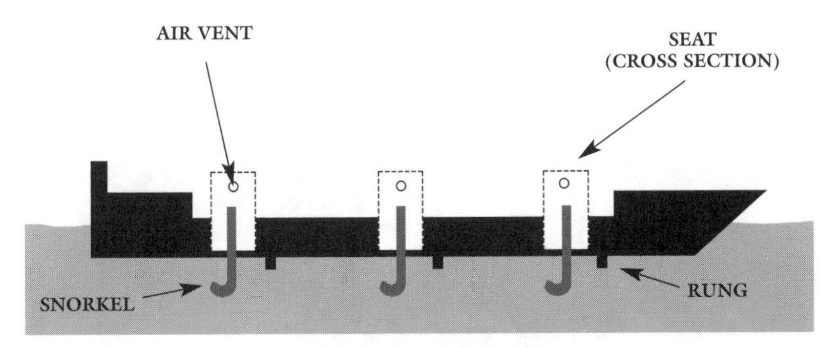

RUNG AND SNORKEL SYSTEM

Hang Glider

Gliders were first used by commandos in World War II, with the British, Americans, and Germans all getting into the act. Today, hang gliders are more likely to be used because they are smaller, easily hidden, simple to operate, and often unexpected. You need an elevated launching point.

Hang gliders must only be used by guerrillas at night, and the operator must be an expert at aerial navigation via terrain association or, if available, Global Positioning System (GPS). He must stay clear of the moon and stay sufficiently high until time to land so that he is not heard; yes, all hang gliders make some noise as the wind pushes against the fabric and frame of the glider.

Bicycle

Guerrillas used bicycles extensively during World War II, and the Vietcong also used them to move supplies and weapons not only along the Ho Chi Minh Trail but off it as well.

Some conventional forces are currently experimenting with mountain bikes as a means for messengers and even reconnaissance units to get about, and the guerrillas can do the same. These rugged bikes can also move supplies.

Animals

Horses and other animals can be used, too. A neat trick I once used to move some sheet explosives into an area was to sew the explosives right into the saddle blanket of the horse. No one suspected a thing.

Pack animals may take the place of vehicles if that is all you have.

Now let's look at some mines and booby traps.

CHAPTER 8

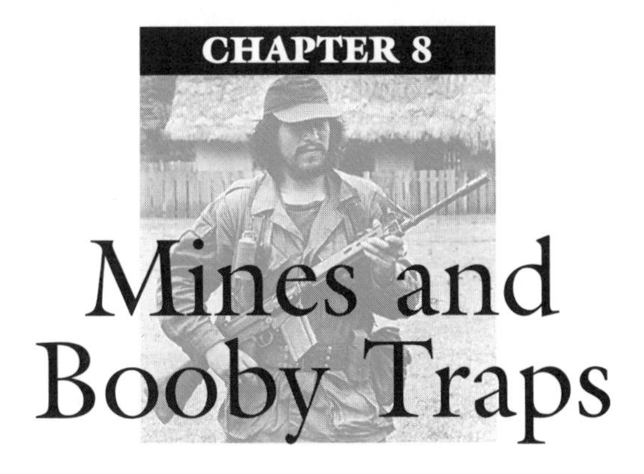

Mines and Booby Traps

"The mine issues no official communique."

—Admiral William Pratt, 1942

Mines and booby traps are to the guerrilla as lies and half truths are to politicians and many journalists; they go hand in hand and play a critical role in conducting day-to-day business. The guerrilla force that does not use mines and booby traps as a combat multiplier stands a substantially reduced chance of victory.

Mines and booby traps first came into their own on a grand scale in World War II when antitank and antipersonnel mines were used heavily by both the Axis and Allied powers; booby traps were used but to a much lesser degree. But it wasn't until the Vietnam War that we really began to see just how devastating an effect these guileful weapons could have on an army. Their surreptitious nature and tendency to brutally maim rather than kill outright makes them particularly effective in lowering the enemy's morale and shattering not only their will to fight, but that of their country, too. Also, they are readily available in the world arms market and can be very affordable, with some antipersonnel mines costing as little as $2 each. When you add this to the fact that booby traps are just as easy to manufacture and rig as antipersonnel mines are to purchase, you begin to see why they are so important to the guerrilla.

Every mine or booby trap set where a guerrilla might accidentally detonate it because he didn't know it was there must be marked to prevent this. A system of apparently meaningless marks (meaningless to the enemy, and perhaps even unnoticed) must be developed that warn any guerrillas passing by that a mine or booby trap lies ahead. These marks might be something as innocent looking as a machete slash against a tree.

Mines are divided into three categories: antitank or antivehicle, antiship, and antipersonnel. The judicious use of each can substantially turn the tide in a guerrilla war, and if the guerrillas have a reliable source of resupply, their combat power is all the greater.

DETONATION SYSTEMS

Mines and booby traps can be detonated by a number of means, including the following:

- pressure
- pressure-release
- tension-release
- on-command
- magnetism
- timer
- antidisturbance
- air pressure change
- air temperature change

Which detonation mechanism the guerrilla selects will depend on the circumstances.

Pressure

Pressure-detonated mines are among the most common types found in guerrilla wars. They work on the simple principle

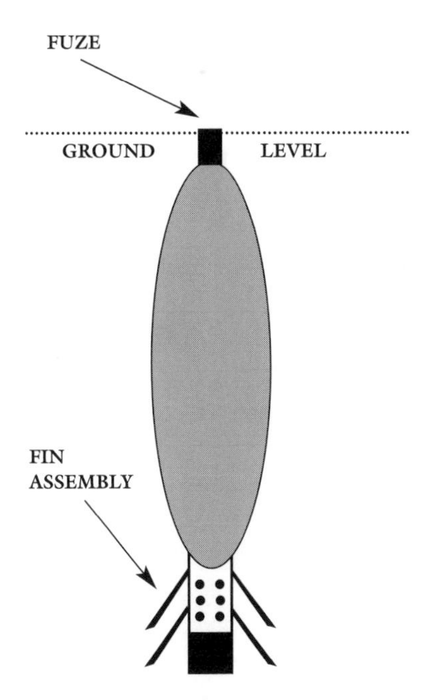

MORTAR ROUND MINE

of someone or something applying pressure to the top of the mine, which detonates it instantly in most cases. They are commonly placed on trails, paths, and roads. Caution must be taken to ensure that it is placed at a depth that is just right for detonation.

Some types of mortar rounds can be rigged as mines by placing them nose up in the ground with the bore riding safety pin removed and setting the fuse assembly on impact. Failure to remove the bore riding safety pin will result in the round's failing to go off.

Pressure-Release

The pressure-release mine is exactly the opposite of the pressure mine. This mine requires that the pressure keeping the mine from detonating be released or reduced in order for the mine to detonate. Such mines often rely upon some ruse perpetrated upon a careless or unsuspecting enemy soldier where the soldier picks up or moves an object sitting on the mine's pressure plate or switch. A classic example of this is when a guerrilla force leaves something that the enemy would find of interest on top of the unseen mine—a weapon, map case, radio, stack of papers, or what have you. When the enemy soldier removes it—whump. Even more insidious is placing a soda or beer can upright along a road or path, underneath which is a pressure-release mine. Many soldiers just can't resist kicking at cans, so . . .

Pressure-release mines can also be placed under the dead bodies of guerrillas and enemy soldiers.

Tension-Release

Tension-release detonators are normally associated with booby traps. These are the classic "trip wire" devices that cost so many American GIs body parts and their lives in Vietnam.

When setting these booby traps it is important and very wise to have two sources of tension on the wire. This adds an extra dimension of safety to the operation.

The wire can be run across trails at various heights, and I recommend using the 5-to-1 rule. This rule says that after every five trip wires set at ankle height, the next one should be set at head or shoulder height. You do this because if the enemy point man finds five wires at ankle height he may assume that they are all set at ankle height. Hopefully, he will be proven fatally wrong.

On-Command

The on-command (command-detonated) mine is most often a claymore, a container of pellets backed with an explosive charge that hurls the hundreds of pellets toward the enemy. It is a devastating weapon that can cripple or kill numerous enemy in a single well-directed blast. An electrical charge detonates the mine.

Magnetism

Magnetic mines are usually reserved for antishipping purposes. They can be set on the bottom of a harbor, and when a ship passes over them the magnetic pull lifts them off the bottom and pulls them against the ship's hull for immediate detonation.

Timer

Mines of many styles can be rigged with timers in the factory or by the guerrilla in the field. Limpet-style mines are commonly used by special operations forces such as Force Recon Marines, Navy SEALs, and the British Special Boat Squadron to sink enemy ships. If a guerrilla force can come by such mines and use them against enemy shipping, they have done good and added a new dimension to the war.

Antidisturbance

Both mines and booby traps can be rigged with antidisturbance systems designed to detonate the mine when someone tampers with it. A simple mercury switch is all that is really needed, and these can be fashioned with the insides of a standard household thermostat containing a vial of mercury, which allows an electrical circuit to be completed.

Air Pressure Change

A more sophisticated detonation device uses a barometer to explode the mine or booby trap when the air pressure rises or falls to a certain level. This is a good system to use when your intelligence says an enemy unit will be moving into a certain area ahead of or behind a storm front.

Air Temperature Change

The same principle as the air pressure change detonator can be used to rig a device to go off when the air temperature reaches a certain level. A digital thermometer is required.

Mines and booby traps are one of the guerrilla's best friends. Use them wisely and watch the morale of the enemy plummet.

Above: Unexploded antipersonnel bomblets from a cluster bomb unit (CBU) such as this can make excellent booby traps.

Right: Comrades of this Tupac Amaru terrorist booby-trapped the Japanese ambassador's home in Lima after seizing it and hundreds of hostages in December of 1996. Fortunately their crimes failed when a CIA Schweizer RG-8A reconnaissance aircraft overflew the premises and pinpointed the booby traps as well as the exact location of everyone inside the mansion. This allowed the Peruvian commandos to take the building down with the loss of only two of their men (the unit's ops officer and a lieutenant) and one hostage (who apparently died of a heart attack). All of the terrorists were correctly killed. The moral of the story may be that the guerrilla should never turn to terrorism.

CHAPTER 9

Ambushes

"We have the power to knock any society out of the 20th century."
— Defense Secretary Robert McNamara, 1964

Just hearing the word "ambush" can make a guerrilla fighter's jaw tighten and eyes squint, for if there is one hallmark of a guerrilla war, it is the ambush. The ambush is probably the oldest tactic in the guerrilla's warfighting manual, and the reason it is still there is simple: it works, and it works remarkably well when planned and executed correctly.

The list of guerrilla forces who used ambushes is long and impressive and includes dozens of American Indian tribes, the Huns, the Mongols, the Swiss, the Afghans, the French, Americans in nearly every war they have fought, the Russians and the Soviets, the Japanese, the Chinese, the Vietnamese, the Tamils, and thousands more.

The guerrilla force must be expert at setting and executing ambushes in a variety of situations. Once the guerrilla attains the ability to ambush enemy forces with bloody results, his combat power is greatly multiplied.

WHY AN AMBUSH?

Besides the obvious overall goal of weakening the enemy physically, tactically, spiritually, and logistically, the ambush is meant to either destroy a certain enemy force (and reap the many rewards that come from that destruction) or harass a certain force in order to wear it down and reduce its combat power.

When an enemy force is destroyed, meaning it can no longer fight as a unit, the guerrillas can benefit by capturing the following:

- personnel
- weapons
- equipment
- intelligence information

They will also benefit from the following:

- a rise in their morale
- increased ambush experience
- increased general combat leadership experience

The guerrillas benefit in the following ways when they successfully harass an enemy force:

- by forcing the enemy to abandon or alter their plans
- by keeping the enemy on the defensive
- by lowering the enemy's morale and will to fight
- by demonstrating to the civilian populace that the enemy is powerless against the guerrillas

And regardless of the type of ambush—destruction or harassment—the enemy will become less aggressive and more uneasy and distressed when operating in guerrilla country. This will result in the enemy's substituting caution for aggression, which orients them more to the defense than the offense, and that is precisely where the guerrilla wants his enemy.

AMBUSH CLASSIFICATIONS

The type of ambush you conduct will depend entirely on the tactical situation at the moment. There are two types of ambushes: deliberate and hasty.

Deliberate

The deliberate ambush is an ambush planned against a preselected unit at a predetermined time and place. Solid intelligence and detailed planning are the two elements that make a deliberate ambush possible and worthwhile.

Deliberate ambushes require information on the following:

- the size of the enemy unit (squad, platoon, company, etc.)
- the composition of the enemy unit (infantry, mechanized infantry, motor transport, headquarters, supply, etc.)
- the disposition of the enemy unit (casual, somewhat alert and defensive, very alert and defensive)
- the strength of the enemy unit (weapons, leadership, support equipment, etc.)
- the route of march of the enemy unit and estimated time of the enemy's arrival

A deliberate ambush could easily result in the disabling of an important weapon system, which the guerrillas can then use.

Female New People's Army terrorists learn hand-to-hand combat, skills that can be used in an ambush designed to take hostages.

Hasty

The hasty ambush is conducted quickly on targets of opportunity. To make it work the guerrillas must think and act quickly and have set SOPs that make them react correctly when the time comes.

Hasty ambushes require the following information:

- the approximate size of the enemy unit
- the suspected composition of the enemy unit
- the suspected disposition of the enemy unit
- the approximate strength of the enemy unit
- the current route of march
- the estimated time of arrival in the kill zone

Stream crossings can be excellent ambush sites if the enemy fails to take security precautions.

Support vehicles traveling along roadways where the trees grow right up to the road are frequently easy prey.

AMBUSH ESSENTIALS

The conduct of an effective ambush requires three essential components.

Surprise

It is no surprise that the offensive fundamental of surprise is critical to the successful conduct of an ambush. From surprise, all else comes when dealing with ambushes, for without surprise, you have no ambush. Good intelligence, thorough planning and rehearsals, individual and team preparation, and masterful execution are what lead to surprise.

Fire Coordination

This is frequently one of the most difficult and time-consuming factors to master. Communication and attention to detail are the keys to solid fire coordination, without which the ambush is doomed to failure. Every man must know precisely what to do and when to do it if the kill zone is to be filled with the bodies of the enemy. A mistake in fire coordination means a loss of concentrated mass fires in the kill zone.

You will need a weapon capable of stopping the enemy's forward movement immediately.

Command and Control

The ambush is the most intense form of combat and, as such, requires the very best in command and control. Leaders all up and down the chain of command must have a tight grasp of the procedures that control the ambush, from the insertion to the setting up to the execution to the extraction. SOPs, thorough planning and rehearsals, and supervision are what it takes to command and control an ambush effectively.

Command and Control Goals

The guerrillas must use command and control measures to do the following:

- detect the enemy's approach and inform the ambush team
- allow the enemy to fully enter the kill zone before initiating fire
- utilize immediate action if the ambush is discovered prior to initiation
- utilize direct and indirect fires when needed
- utilize search and sweep techniques immediately after the ambush
- safely and tactically withdraw once the ambush is finished

AMBUSH FORMATIONS

The type of ambush formation you select will depend upon the following:

- the size and formation of the unit being attacked, as well as its composition and disposition
- the weaponry being carried by the unit being attacked
- the terrain
- the vegetation
- the size, composition, and disposition of the ambush force
- the weaponry available to the ambush force

Once the guerrilla leader has determined these factors he can better select the best ambush formation to use.

Linear (Line) Ambush

The linear ambush situates the ambush force parallel to the enemy force. After pinning the enemy in position by sealing off both ends of the long axis with mines or other weapons, the ambush force attacks the enemy with large volumes of fire along the enemy's flank that is exposed to the ambush force.

The guerrillas must use caution to ensure that the majority of enemy troops and weapons is within the kill zone; an enemy unit that is strung out on its long axis is unlikely to be damaged seriously by a linear ambush.

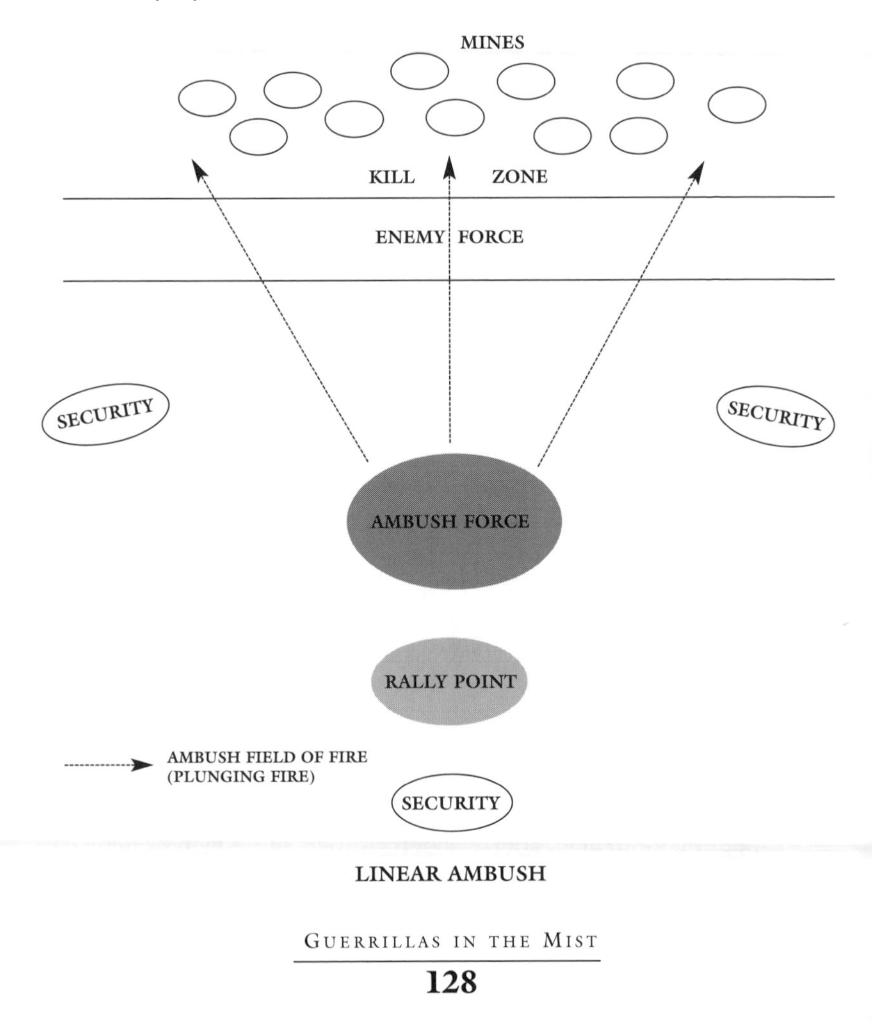

LINEAR AMBUSH

Box (Rectangle) Ambush

The box ambush is especially effective against a force that has little or no experience with ambushes. It relies upon four ambush teams forming the four corners of a box with the enemy inside the box. One corner engages the enemy at a time. When the enemy turn to fight in that direction, the initiating team ceases fire and appears to be withdrawing. At that time the team diagonal to the first team opens fire and forces the enemy to shift their focus of attention. As they do so, a third team engages the enemy as the second team ceases fire and appears to withdraw. When the enemy go after the third team, the third team also ceases fire and appears to withdraw. Now the fourth team opens fire to finish off what's left of the enemy force.

Pyramid (Triangle) Ambush

There are two types of pyramid ambushes: tight configuration and loose configuration.

The tight pyramid or triangle is used with interlocking fields of fire from automatic weapons set at the points of the configuration when the direction from which the enemy will be approaching is unknown. This configuration is fairly secure but requires at least 20 or so men due to the risk of the enemy assaulting through.

Similar in concept to the box, the loose triangle has the automatic weapons engaging the enemy's lead first until the enemy tries to assault through them. When they do so, one of the flanking automatic weapons opens fire to draw the enemy off. When the enemy tries to attack that team, the other flanking team engages them.

An offshoot of this is to have assault teams standing by to close with and destroy the enemy as soon as they are in disarray and being worn down by the automatic weapons fire.

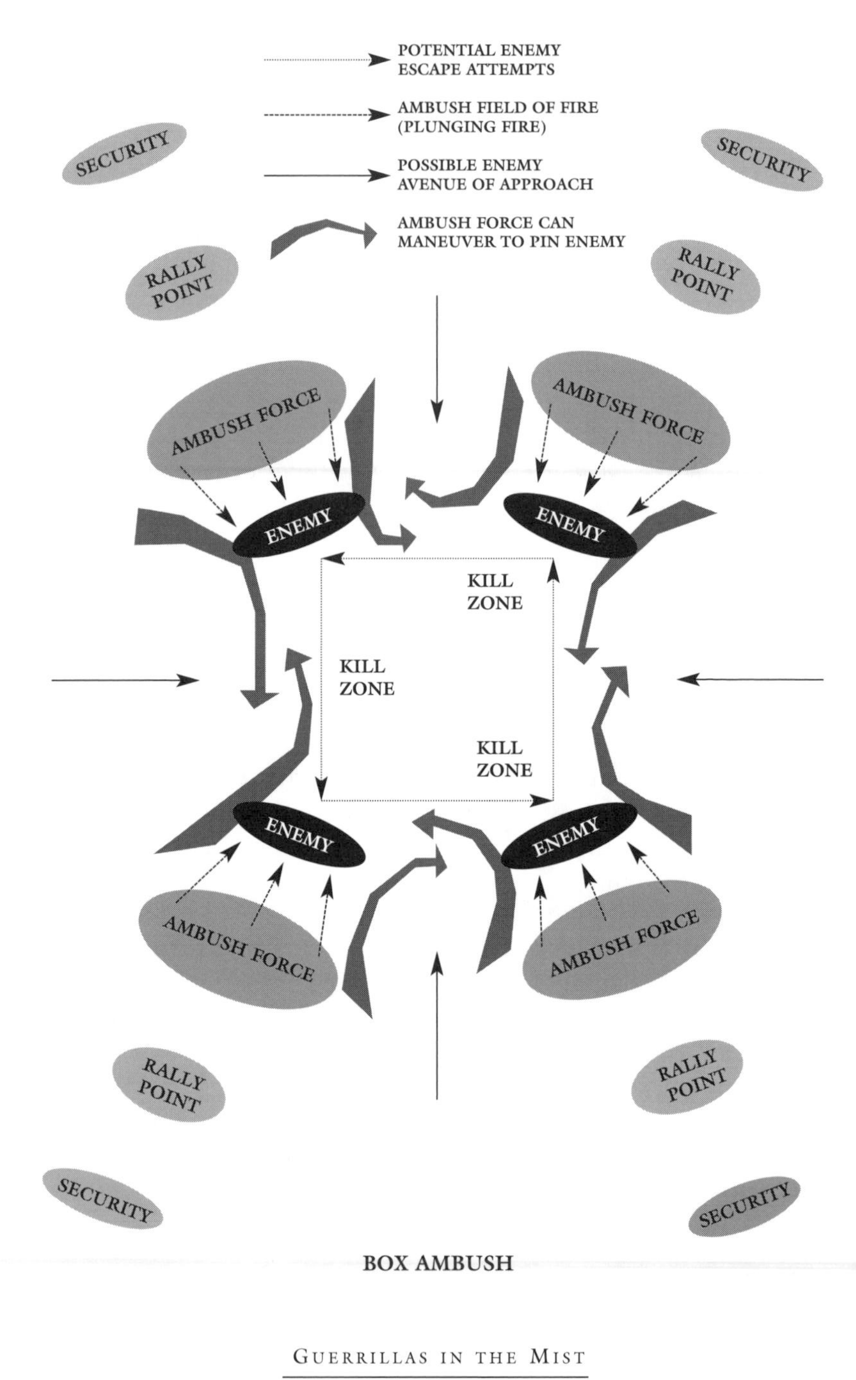
POTENTIAL ENEMY
ESCAPE ATTEMPTS
AMBUSH FIELD OF FIRE
(PLUNGING FIRE)
POSSIBLE ENEMY
AVENUE OF APPROACH
AMBUSH FORCE CAN
MANEUVER TO PIN ENEMY
SECURITY
SECURITY
RALLY POINT
RALLY POINT
AMBUSH FORCE
AMBUSH FORCE
ENEMY
ENEMY
KILL ZONE
KILL ZONE
KILL ZONE
ENEMY
ENEMY
AMBUSH FORCE
AMBUSH FORCE
RALLY POINT
RALLY POINT
SECURITY
SECURITY

BOX AMBUSH

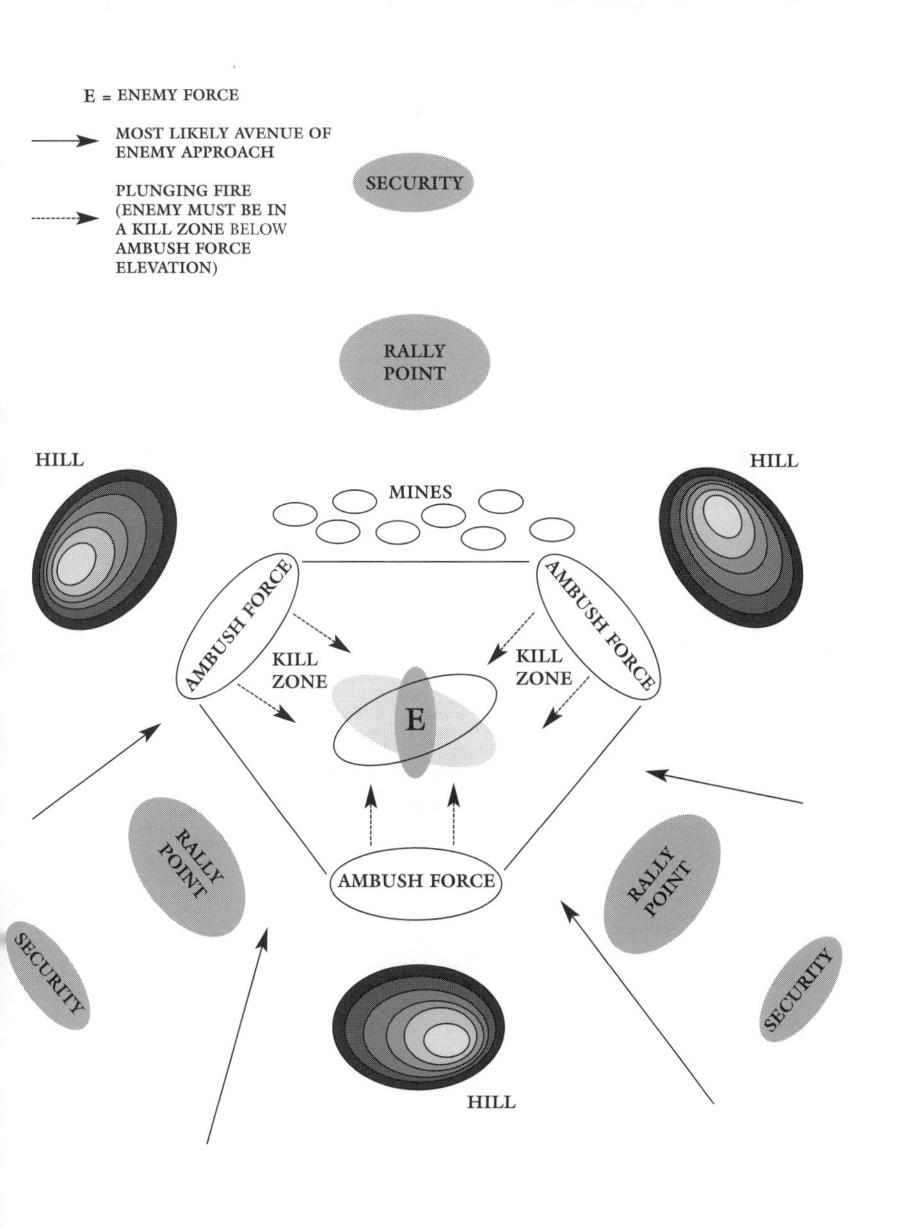

PYRAMID AMBUSH

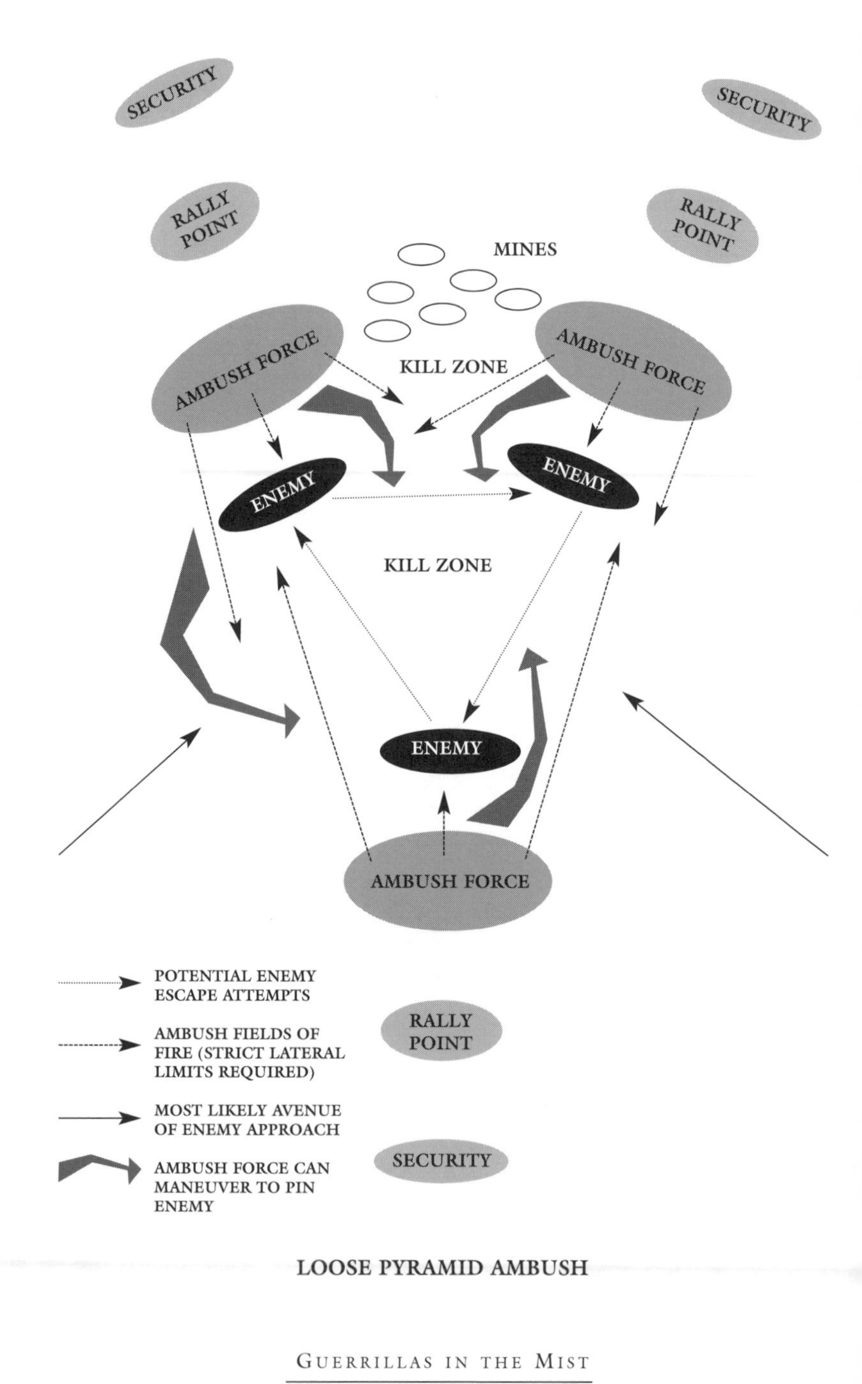

LOOSE PYRAMID AMBUSH

Z Ambush

When an ambush force may be faced with an exposed flank, a reinforcing enemy unit, a single envelopment, or even a kill zone with a potential escape route, the Z ambush is used. It provides a secondary kill zone in case the enemy manages to get through the parallel ambush team.

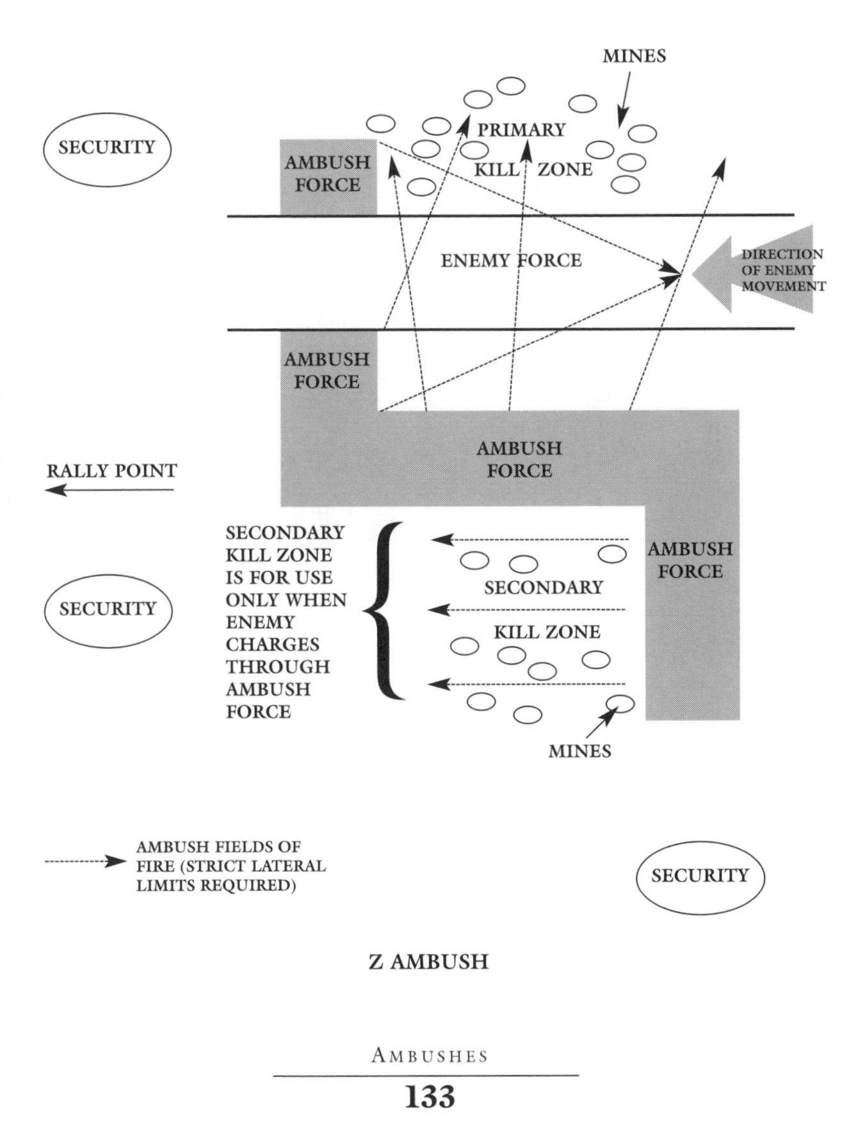

Z AMBUSH

L Ambush

The L ambush is considered the classic ambush because it can be used in various situations. The team occupying the leg parallel to the kill zone delivers flanking fire that supports the blocking team's fire into the enemy element's lead. The ambush force must ensure that the enemy's rear does not escape. This can be accomplished by setting a claymore mine to cover this possibility.

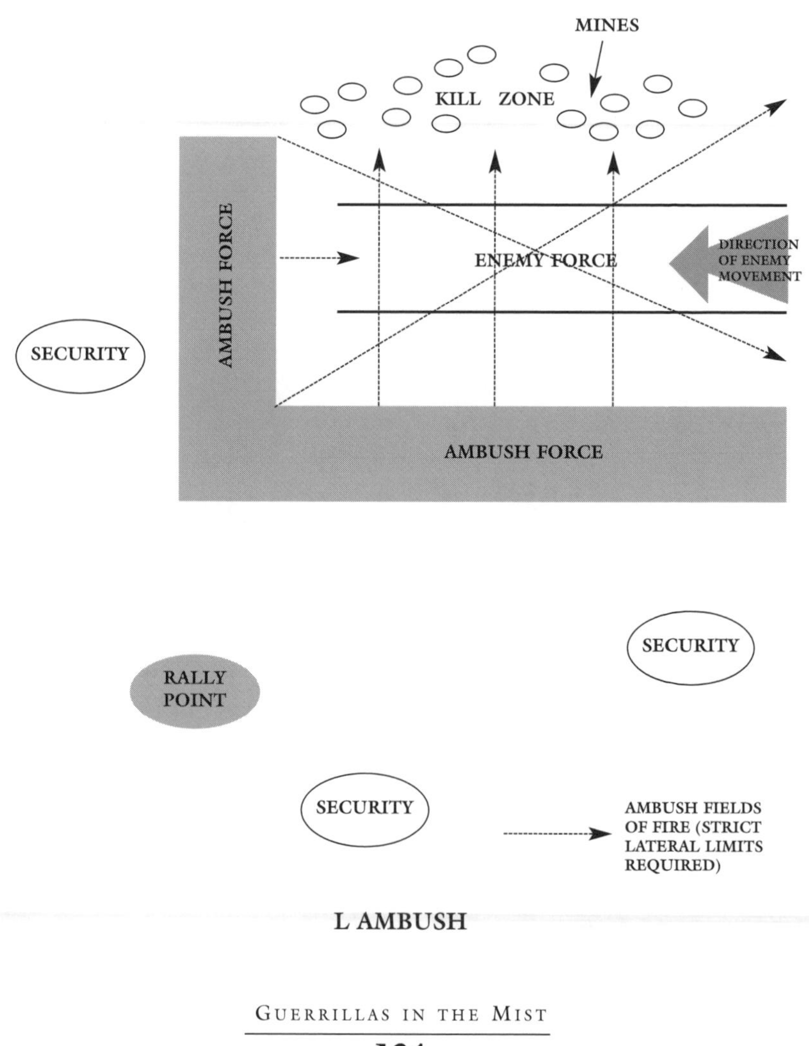

L AMBUSH

T Ambush

This formation takes its name from the shape the opposing forces take in the ambush. The ambush force is perpendicular to the route of march of the enemy force and is able to deliver harassing fires onto the enemy's lead as well as down their long axis in a conical fashion. It can also be used when the direction of the enemy's approach is somewhat in doubt but believed to be most likely from a right angle.

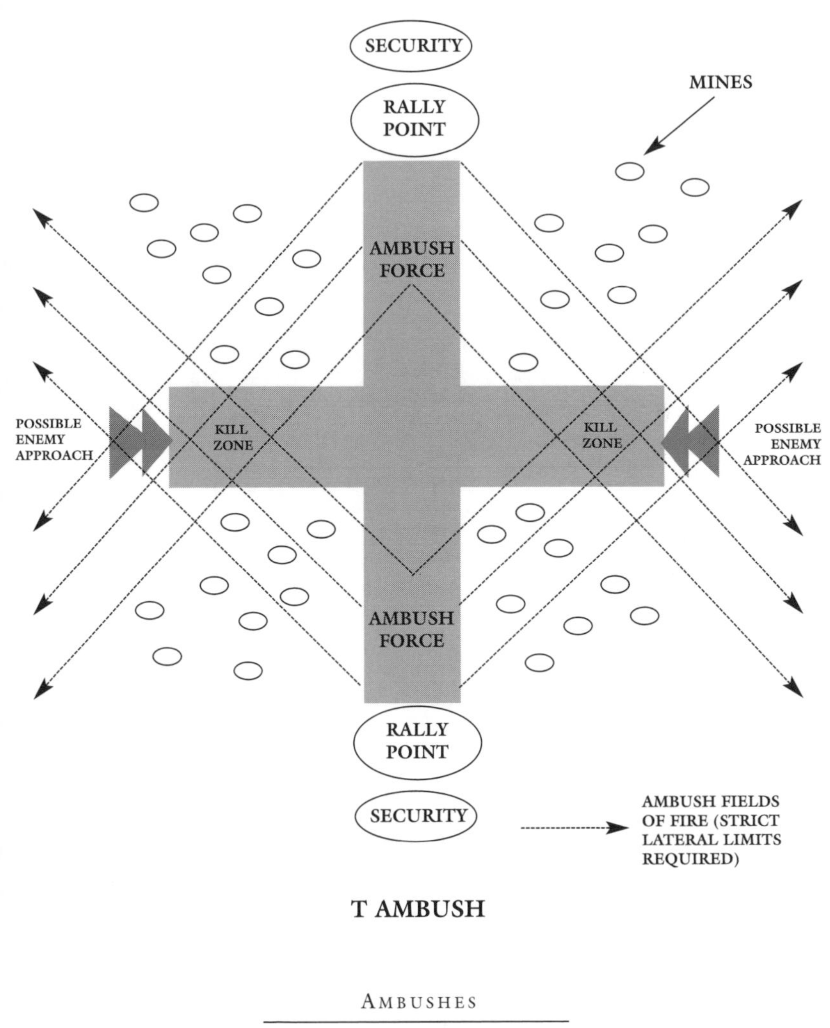

T AMBUSH

V Ambush

This ambush is also quite versatile and gives the ambush force an excellent chance of complete surprise when the ambush is set in a wooded area along a trail. Strict fire discipline is needed to prevent one leg from shifting fire too far toward the opposite leg. The diamond-shaped kill zone is covered with interlocking and enfilading fires.

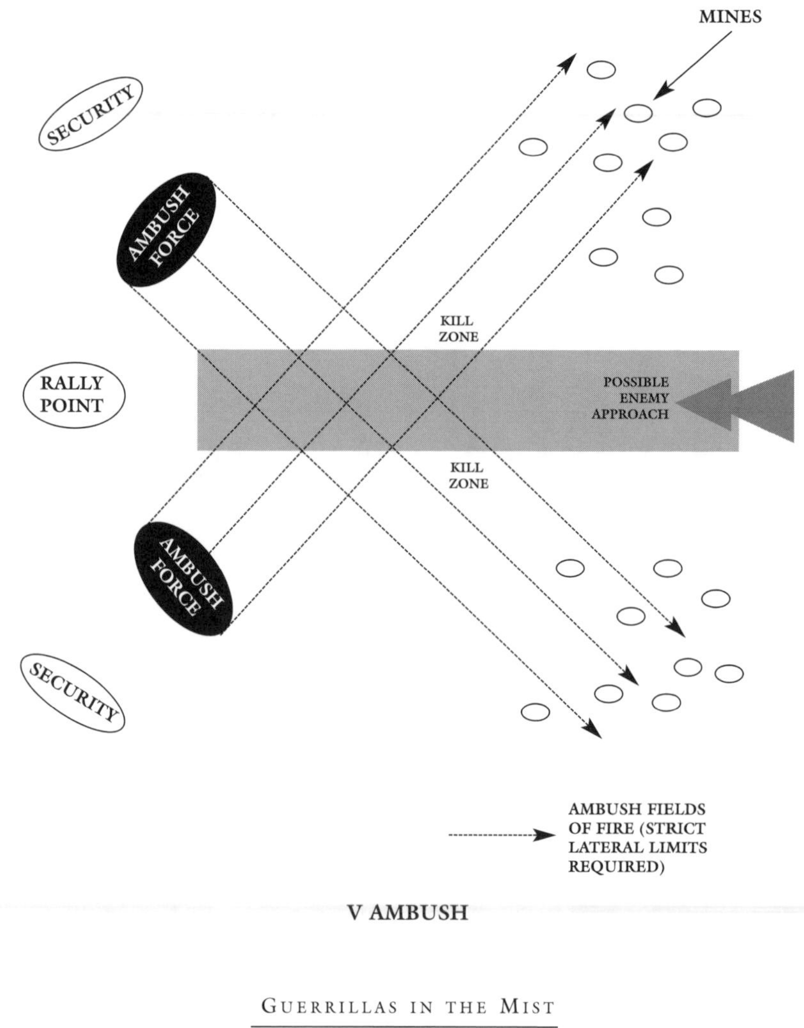

V AMBUSH

SPECIAL AMBUSHES

Special ambushes are limited only by the situation and the audacity and imagination of the guerrillas.

Helicopter Ambush

By disabling an enemy helicopter in a kill zone but leaving some of the enemy alive inside or just outside the chopper, the guerrillas can lure in another helicopter or helicopters in an attempt to extract the downed bird's crew and/or troops. To do this the guerrillas simply cease fire and remain hidden until the rescue choppers arrive.

In many cases the rescue force will first fly a couple of gunships along the perimeter of the landing zone to check for and clear guerrilla forces. This makes camouflage and discipline critical: no guerrilla must be seen or open fire until the rescue birds are on the ground. Then they open with a large volume of concentrated fire to cripple or destroy the other helos. Withdrawal must be immediate if the enemy has the resources to engage the guerrillas from the air with gunships or with supporting arms like artillery and mortars.

Demolition Ambush

A demolition ambush is initiated by first attacking something important to the enemy with explosives. Harassing fire is then delivered to the target as the enemy reacts to the explosion. It may behoove the guerrillas to hold their fire for a while before opening up with their automatic weapons from a distance; the enemy will likely be expecting some immediate harassing fire and by delaying for a short while the guerrillas may find that the enemy, once it appears to them that no harassing fire is imminent, relaxes their guard somewhat.

Claymore-Sniper Ambush

If the enemy is moving across fairly open ground with little cover, a claymore mine can be detonated to stop his forward progress. Several snipers are positioned around the kill zone and withhold fire until the enemy thinks that the guerrillas have fled. Once the enemy comes out into the open again the guerrilla snipers pick off as many enemy as possible and then withdraw.

Ambushes are one of the guerrillas' best ways to even the sides.

CHAPTER 10

Prisoner Interrogation, Exploitation, and Indoctrination

"Death is lighter than a feather, duty heavy as a mountain."
—Emperor Meiji, 1883

For a few years I was an instructor at the U.S. Navy SERE School in Maine. One of my jobs at this most demanding and challenging school was within the Resistance Training Laboratory (RTL) training and testing students on their ability to survive a simulated prisoner of war (POW) experience honorably. Also during this time, I served as an advanced SERE instructor teaching graduates of the basic SERE course things they needed to know prior to making a deployment overseas in the near future. In both jobs I worked for former prisoners of war who had been shot down over North Vietnam and captured in 1968. Their experiences helped me—and the rest of the instructor staff—better understand the myriad factors governing the interrogation, indoctrination, and exploitation of prisoners of war.

Immediately after finishing my tour at the school I was sent to Saudi Arabia and then Kuwait with a Marine rifle company for the Gulf War, from the buildup to the final withdrawal from Kuwait in May of 1991 of the last Marine combat unit, which happened to be my unit (2nd Battalion, 4th Marines; then-

Prisoners are very valuable commodities. Exploit them properly. (Department of Defense photo.)

Lieutenant Colonel Kevin A. Conry, commanding). I found that, just as I expected, it is often easy to make prisoners useful in some way, sometimes extremely useful.

CAPTURE AND HANDLING

Let me make this perfectly clear—a prisoner of war is a valuable commodity that you, the guerrilla, must learn to exploit. A very valuable commodity. As such, you must go to the greatest lengths to make extremely sure your men do not—inadvertently or otherwise—injure, maim, harm, or otherwise damage the prisoner.

Combat is an incredibly emotional experience. All at once it can be terrifying, thrilling, shocking, horrifying, disgusting, repulsive, maddening, sickening, humorous, confusing, and mind-numbing, and every guerrilla is going to be, to some degree, affected by these emotions. This undeniable fact is what you, as the guerrilla commander, are going to have to deal with on probably a regular basis when it comes to prisoners of war and how they are handled during and immediately after their capture and, if you are required to hang on to the prisoners for an extended period, perhaps for months on end after their capture. Emotions—those of the guerrillas—are the greatest threat to the value of the prisoner of war, and you must do everything in your power to prevent emotional responses from degrading the value of your prize.

Prisoners of war are afforded certain rights under the Geneva Conventions Relevant to the Treatment of Prisoners of War, which was signed in August of 1949 by many signatories and has since been signed by many more. Abiding by these conventions actually increases the combat power of the guerrillas in many (most) circumstances, so it behooves the guerrillas to abide by them. Brutality for brutality's sake is of almost no value to the guerrillas and only serves to gain immediate physical compliance. Although there is something to be said for physical compliance, there are other, more valuable means for gaining such compliance, as we will see.

The 5 Ss

The American military uses the 5 Ss—search, silence, segregate, speed, safeguard—to lay out how prisoners of war should be handled initially on the battlefield and all the way to their permanent detention facility. By sticking to this formula, the guerrillas will reduce the number of problems they have between the time they actually capture the prisoner and the time they turn him over to higher command.

Search

The search is the first thing the guerrilla does upon capturing an enemy soldier. After taking the prisoner's obvious weapons away, the guerrilla conducts an immediate search of the enemy soldier's person, including his uniform. He thoroughly searches pockets, pack, and other web gear, as well as boots, hat, gloves, and every other uniform item for weapons or anything that might be used as a weapon (a comb, pencil or pen, keys, watch, compass, belt, boot laces, identification tags and their chain, and so on). Once the tactical situation permits, the guerrilla performs a strip search, and this includes body cavities and depressions. The anus, ear canals, mouth, bottoms of feet, armpits, groin area, small of the back, hair, nostrils, navel, posterior knee, palms of the hands, and other potential hiding spots are checked and cleared.

All searches are done with one man doing the search and one man covering the prisoner with a rifle. The searcher must never step between the prisoner and the man with the rifle, for obvious reasons. If the guerrillas have the funds (and this doesn't require much), some electronic dog collars should be bought and placed on the prisoner during the detailed search. Should the POW decide to get feisty, a guerrilla just pushes a button and a nice stream of electricity flows directly into the prisoner. A stun gun is another worthy investment.

The guerrillas must translate immediately all documents and process them for their intelligence value and their exploitation and interrogation potential. Letters to and from home, warning and operations orders, radio frequencies, SOP cards, resupply requests, and much more may be of substantial value to the guerrilla effort.

Silence

Every POW must be silenced from the moment of capture until he is well away from the battle area and in a controlled situation. Masking tape is excellent for this (yes, it hurts when you take it off but it does not harm the prisoner). Make sure he can breathe through his nostrils when you put the tape over his mouth. A friend of mine once lost a prisoner (read: the prisoner suffocated) after taping his mouth shut. The guy had a cold and his congested nose didn't allow enough air into his lungs, so . . .

Depending on the tactical situation, you may have to bind the prisoner's legs and arms as well to keep him from thrashing around.

Segregate

It is permissible to segregate prisoners by rank upon their capture so as to prevent immediate escape attempts, and escape attempts are more likely to succeed when they are undertaken soon after capture, before the prisoners are deposited in a more permanent detention facility. In the first 24 hours or so after capture, the new prisoner usually has the following points in his favor:

- He is close to his unit or friendlies.
- He is often in excellent physical condition.
- He is in situations such as traveling in vehicles that make escape easier.
- He is in situations that otherwise make escape easier because of the fog of war.

By segregating prisoners by rank (junior enlisteds, NCOs, and officers), in many cases the guerrillas can undermine the organization of covert escape committees and committees that facilitate escape in some way.

Later, when the prisoners are in more permanent facilities, it may behoove the guerrillas to segregate prisoners by race, religion, gender, and even age. Such segregation can make exploitation easier and more productive.

Speed

Removing the prisoner from the battle area as quickly as possible is crucial to his future exploitation. By getting him away from the fighting and speeding him to a safe area, the guerrillas are more likely to prevent his successful escape or additional injury.

But with speed comes the danger of overlooking a security concern. Every precaution must be taken to deny the prisoner the opportunity to slip away unseen or make a bold dash for freedom in the confusion of the fight. Plans and SOPs must be carefully laid to increase the chances of a successful extraction of the prisoner.

Safeguard

Every prisoner must be afforded sufficient protection from enemy fire and friendly retaliation. The more damage you do or allow done to a prisoner, the more likely he is to be of little use. Remember that the prisoner of war is a most valuable commodity that is to be zealously safeguarded regardless of the tactical situation. If this means that he must be allowed to keep his personal protective gear like a helmet, gas mask, and flak jacket, then so be it.

The guerrillas must move the prisoners frequently to help prevent rescue attempts by their comrades. They must conduct searches continually so that they can detect and deal with secret stashes and covert organizations within the ranks of the prisoners. If facilities permit, prisoners should be kept in solitary confinement and secured with redundant security systems.

Interrogation and Exploitation

Prisoner interrogation is an art unto itself and is covered in its entirety in the Paladin Press book *Make 'Em Talk: Principles of Military Interrogation.* Given this, I won't cover all the approaches and tricks of the trade here but will instead give you general advice that will assist you in interrogations.

Alcohol can often be used to loosen a prisoner's tongue faster than any other approach. This Spanish Legionnaire knows the deal.

Torture Is Out

Forget torture altogether. This surprises you? It surprised me, too, when I was training to be an interrogator under the tutelage of career Marine interrogator Gunnery Sergeant George Misko (who conducted many interesting interrogations during the Gulf War, and elsewhere, a man who can terrify a prisoner with a mere look), former prisoners of war Cdr. Bob Fant USN (Ret.) and Cdr. Tim Sullivan USN (Ret.), and now-retired Navy SEAL Master Chief Tom Keith, the latter of whom conducted numerous field interrogations in the Mekong Delta.

Torture gains compliance, physical compliance, but it doesn't often elicit useful information that is reliable, and it doesn't serve any purpose in the practical exploitation of the prisoner. For instance, during an exploitation session in Hanoi where several American POWs were brought before a select group of media (selected by the Communist regime, that is), one prisoner of war was forced to walk on stage in front of the cameras and was then told to bow. The prisoner intentionally walked out and acted like a zombie, then bowed stiffly to the assembly first to the front, and then at 90-degree angles to "box the compass." This made him appear like an automaton who had lost his mind, and it was made even worse for the man's captors when a guard ordered him to bow again. This gave the clever prisoner another chance to box the compass with bows, three of which went toward the sides and rear of the stage.

During another exploitation session in Hanoi, a group of American Communist media was allowed to question several POWs. A bowl of fruit had been placed on the table to give the idea that the prisoners were being fed well, which was far from the truth, of course, but one prisoner saw an opportunity to turn the tables on the turncoats by snatching a large handful of the fruit and feeding ravenously as they questioned him, showing that he was underfed.

The very best way to extract useful information is to use proven approaches (outlined in detail in *Make 'Em Talk*) that do not call for genuine physical abuse or torture. It is almost always easier to get useful and accurate information by using trickery.

Professionals Only

The interrogation of prisoners by guerrillas other than those trained in interrogation techniques must be avoided. There is much too much to be lost at the hands of an irate, untrained interrogator in the field. Discipline and extraction SOPs will help get the prisoner away from the battlefield and into the hands of someone who knows exactly what he is doing when it comes to extracting information.

Information Storage and Exploitation Systems

Information gleaned from prisoners must be kept in a safe place and in a system that is easily and accurately accessed by the guerrillas' intelligence network. Systems with holes that allow good information to be lost or not fully exploited must be tended to.

The dissemination of intelligence from the interrogation center down to the troops must be smooth and efficient at all levels. Otherwise, the guerrilla movement is simply spinning its wheels in this area.

Smoke tells of a guerrilla strike made successful by the information extracted from a prisoner by a professional interrogator.

Indoctrination

It may or may not be worth the time and effort of the guerrillas to run an indoctrination program; it all depends on their resources, abilities along these lines, and the education level and resistance techniques of the prisoners. If the guerrillas have the financial and logistical resources, as well as indoctrinators who are experts at their craft, and if the prisoners' educational level is fairly low and they have little or no training in resistance techniques, it is advisable for the guerrillas to indoctrinate the prisoners into their way of thinking. This not only makes them more compliant, it also makes them excellent propaganda sources, and that propaganda can be directed at both the local populace and the enemy's homeland.

But as the previous two examples of exploitation attempts demonstrate, exploitation and indoctrination are not without risk, especially when the prisoners are educated and have been trained to resist enemy indoctrination attempts. Interrogators should always try to learn the prisoner's educational level.

To indoctrinate prisoners—make them think like you think and see things as you see them—the guerrilla must use a smattering of truths mixed with half-truths, lies, disinformation, and vagary to sway them. Propaganda used against them must be of the highest quality; anything less will be seen for what it is: a trick. Persistence and the ability to use seemingly insignificant data as tools in the indoctrination system are two of the guerrillas' most valuable assets.

On the Other Hand

Lastly in this chapter, a word on guerrillas who are captured. Every effort must be made to rescue guerrillas who are taken prisoner, and every guerrilla must be taught escape procedures and how to resist enemy interrogation, exploitation, and indoctrination techniques. Strict accountability must be maintained so that guerrillas taken prisoner are not willingly left behind in the name of political expediency, such as when America left nearly 1,000 American POWs in North Korea and China after the

Korean War during Harry Truman's presidency, and when, as some allege, hundreds more American POWs were knowingly left behind in North Vietnam after the truce was signed in 1973 during the Nixon presidency.

The leadership trait of loyalty must be demonstrated at all levels.

CHAPTER 11

Outthinking the Counterguerrilla Force

"Better to die than be a coward."

—Gurkha saying

Guerrilla warfare is a war of tactics, strategy, operational art, and firepower, but it is also a war of wills and minds. The guerrilla army that understands the counterguerrilla techniques the enemy uses and moves to thwart those techniques at every turn will be the victor. Even America—a country with extensive experience fighting guerrillas, ranging from the Barbary pirates in the early 19th century to Mohammad Farah Aidid's guerrilla thugs in Mogadishu—continues to be stymied by third world hooligans, this despite the world's most powerful nation having been bloodied in guerrilla wars in Central and South America, Africa, Asia, North America, and Europe. Major powers will always, it seems, be vulnerable to the guerrilla because of the modern nation's arrogance, poor leadership (in the capital and on the battlefield), and remarkable ignorance when it comes to how a guerrilla army can and should be engaged and defeated.

THE ADMISSION

The number one problem conventional armies make when going to war against a guerrilla force is refusing to admit that their ally—the nation with the insurgency/guerrilla problem—is suffering from political, social, and economic problems that have given birth to the very insurgency it is now faced with. Who is to blame for these problems is less relevant than finding and implementing solutions to them or, if that fails, accepting the loss early on so that one can get out.

America lost the Vietnam War because untutored, contemptuous politicians believed they could defeat the Vietcong and North Vietnamese Army with sheer firepower and advanced technology, and these same politicians made the fatal mistake of relying upon military officers who, in most cases, never understood the nature of their opponent and what lengths he was willing to go to in order to win. President Johnson, already one of the most egotistical, crass politicians ever to sit in the Oval Office, accepted the recommendations of generals who told him what they believed he wanted to hear—that massive bombing in the north was the answer. History tells us that although the bombing of Hanoi terrified the civilian populace, the people were never in a position to demand that their government stop the war in the south, and those people did truly see the Americans as barbarians committing war crimes against an innocent society. When Nixon entered office he began secret bombing missions in Laos and Cambodia in an attempt to strike NVA and VC bases there, but the policy failed miserably on the national level, even though B-52 strikes were greatly feared by the enemy. Instead of merely increasing the level of violence and letting it go at that, America should have set out to either remove the impetus of the struggle or, if that proved impossible (and sometimes it will prove impossible), cut its losses and fled Vietnam in 1968. Instead, the war didn't end until 1973, a fact that cost tens of thousands more American lives, and all for naught.

A review of the tactics used by the most successful counterguerrilla forces in Vietnam shows that small units of disciplined, aggressive, crafty soldiers who were willing and allowed to fight the VC on their own terms, were the most successful. When Special Forces outfits started training and arming people like the Montagnards and Hmong to defend themselves using guerrilla tactics and Marine units began their strategic hamlet program along with useful cordon and search techniques, the VC suffered serious setbacks. Add Navy SEALs to the Mekong Delta hunting VC with guerrilla tactics and you start to win. But even these successes would never have been enough to win the Vietnam War because the Communists had an almost inexhaustible supply of young men and resolve, and they knew that America was unwilling to fight for decades and see more and more body bags coming home. The truth of the matter is, some guerrilla wars are unwinnable because of the nature of the insurgency itself. Japan learned it in China, America learned it in Vietnam, and the Soviet Union learned it in Afghanistan.

Once a nation providing foreign internal defense (FID) to a friend admits that its friend has social and political problems that led to the insurgency and moves to correct these problems, it can get on with the war. The guerrillas must be ready for this and ready for the strategy and tactics used by both the nation providing the FID and the host nation itself.

COUNTERING THE FIVE STEPS IN COUNTERGUERRILLA WARFARE: THE MALAYAN INSURGENCY

Many modern nations use the successful British counterinsurgency operation in Malaya (1948–1959), where the British fought and defeated a guerrilla army belonging to the Malayan Communist Party, whom the British referred to as CTs for Communist Terrorists (and they were just that), as a good example of how a nation can conduct a prudent counterguerrilla war. One of the most insightful writings on this topic is Lt. Col. Roland S.N. Mans' essay, "Victory in Malaya." Mans served in

Malaya from 1953–1956, and he both filled combat billets on the front lines as a rifle company commander with the First Battalion of the Queen's Royal Regiment and served as a staff officer on the headquarters staff of the 7th Gurkha Division, one of the most dangerous and effective counterguerrilla forces ever to be mustered.

Lieutenant Colonel Mans cites five factors that are critical to a successful counterguerrilla campaign. The guerrilla leader who is familiar with these stands a much better chance of preventing them from ever taking hold.

A Coordinated Intelligence Network

Without an accurate and active intelligence network that can be used by all in need of that intelligence, the counterguerrilla force is going nowhere. Mans used an outstanding police force of sorts to establish and run this network, and it was highly effective. The British knew a great deal about the CTs' plans, tactics, and strategy and were able to thwart many of the CTs' initiatives because of this.

The guerrillas must strive to prevent a police force (or any other enemy unit) from establishing an intelligence network of any kind. The guerrillas can accomplish this by sniping individual intelligence operatives, ambushing the enemy police patrol before they ever reach their objective, and using propaganda to sway the populace to their side. A fine way to accomplish the latter is to invite the local chieftain to tag along for an ambush of a police patrol coming toward his village. One guerrilla will have on his person a set of bogus orders that direct the police to interrogate and then slaughter all the men in the village after they are forced to watch their women and children being gang raped. Immediately after the ambush, while the guerrillas are searching the bodies, the guerrilla with the orders tucks them into the coat pocket of the senior police officer without being seen by the chieftain, who is being distracted. The guerrilla calls out to the leader that he has found some important-looking papers, and the

Feeding false information into the intelligence system, such as a plan by the guerrillas to hit this nuclear power plant, might cause the enemy to lessen security at a port facility, making this ship vulnerable.

leader and the chieftain approach to see the guerrilla removing those papers from the officer's pocket. He hands them to the leader, who immediately examines them.

Now a little acting. Feigning rage and disgust upon reading them, he thrusts them into the chieftain's hand, who then reads them and sees that the police were coming to slaughter his men and defile their women and children.

This ploy can be reinforced by having a "refugee" from a distant village enter the chieftain's village and tell of how he barely escaped with his life after the police came to his village and massacred everyone therein, women and children included. He also tells of how the guerrillas fought valiantly to defend the village but were also killed, and how he lived only because three bodies fell on top of him during the massacre, shielding him from the bullets and bayonets. He then moves on. Of course, he is no refugee, but actually a guerrilla, and there never was any massacre.

Another ploy, which is ongoing, is feeding disinformation and inaccurate information into the enemy's intelligence network. This should be done from many sources and on all levels so that the enemy use up as much time, energy, and manpower as possible backtracking the bogus information.

If you have a civilian informing on you, he must be killed. However, it must be made to appear as though the police themselves killed him, perhaps because they suspected he was a double agent. His death must show plausible evidence that it was the police (or other intelligence operatives) who murdered him. The word will spread quickly that the police cannot be trusted. And the body or head of a police officer should be shown to the village from which the informant came with the explanation that he was the officer who murdered the informant and that the guerrillas took revenge upon the evil enemy for his crime against the people.

Winning Hearts and Minds

The enemy will try vigorously to "win the hearts and minds" of the people. They'll try to do so by caring for the people with expert medical and dental care, food and water, and much more, and they will try to educate the people to show how the government is right and the guerrillas are wrong. Still, unless the government is mending its ways and has moved to give land to the landless and money to the poor, while holding the wealthy criminal elite accountable for their crimes, the guerrillas can always ask simple questions such as, "Has the government given you the land you were promised? Land that is rightfully yours?" When the answer to pointed questions like these is "no," then you have arguing power.

An excellent way the enemy convince some people to see it their way is by providing the people with food. To counter this move the guerrilla has choices, including ambushing the food convoy before it ever gets to the village it is destined for and lacing the food with a strong but nonfatal poison that makes the people sick. This is done by an agent inside the enemy organization who has access to the food before it goes on the trucks. A week or so after the people eat the food and have recovered from their illness, the guerrillas spread the word that all villagers should use extreme caution when accepting food from the government because a plot was unearthed that called for the poisoning of certain villages by the government.

Integrated Command

The military is going to be required to interface on many levels with civilian officials, and this includes any foreign military assisting the government. The military will try to appease the bureaucrats and officials as often as possible, but will endeavor to always make clear to them that the military is in charge at all times, even if it is trying to make nice.

The guerrillas can do a lot of damage here by finding which officials had a well known run-in with the military that was witnessed by several people. Soon thereafter the official is found dead with a single bullet hole to the back of the head, gagged, and with his hands bound—an obvious execution. Word is spread that the government assassinated him because he wouldn't cooperate fully with them. This method can quickly destroy already shaky alliances between the military and civilians.

Patience and Tenacity

A wise counterguerrilla force will be both patient and tenacious, accepting that the war may take decades to win and accepting that they are indeed in it for the duration. The guerrilla can best counter this by applying Mao's principle of destroying the enemy's national strength.

When the enemy are never allowed to rest and recuperate, are constantly being harassed and maimed and killed, and when their own people back home are being convinced that the war is far too costly and probably unwinnable, they will lose patience and their tenacity will wane.

Training and Aggressiveness

This is the most difficult factor for the guerrillas to counter. The training and aggressive nature of the enemy will likely be instilled in them well before they come after the guerrilla, and once they have that training and aggressive spirit it is troublesome to take it from them.

Nevertheless, the enemy's aggressiveness can be depleted somewhat through the sagacious use of mines, booby traps, and, to a slightly lesser degree, ambushes. (Ambushes are more easily countered in most cases than mines and booby traps.) Everyone fears being maimed more than anything, and the more graphic maimings the enemy suffer, the more cautious they are likely to become. Caution is like a cancer when smeared on aggressiveness.

MANS' PRINCIPLES

Lieutenant Colonel Mans set down four principles that he believed—and with good reason—were the reason the British

Some simple training will have guerrillas poisoning the enemy's water supply with toxic waste.

were successful in Malaya. The guerrillas themselves can apply these same principles in their quest for victory.

Initiative and Aggression

The success of the guerrillas in individual operations will often depend on the initiative and aggressive action taken by NCOs and how well they are able to lead their men in like action. These small unit leaders must do whatever it takes to get their men to be the most physically resilient, hard-charging dogs of war they can possibly be. The guerrillas must all be almost immune to the most extreme hardships over an extended period of time—usually years and sometimes decades. They must live and operate for one thing and one thing only, that being the complete destruction of the enemy no matter where they go and what the conditions there are.

Firearms and Explosives Expertise

Mans stressed the criticality of being an expert marksman at short ranges, but the guerrilla must go beyond that to being highly proficient with all weapons at short and medium ranges, and be at least proficient at ranges up to about 500 yards. He must also have a thorough understanding of explosives and be able to kill silently with a knife.

But Mans also realized that it is impossible for the guerrilla leader to train a unit in which every man has the same marksmanship skills, and he recommended keeping your best shots "up front" where they can do the most good. But everyone must be able to shoot quickly and shoot accurately. Practice is the key to achieving this; there is no other way.

Navigation

Training in land navigation—to include not only traditional map and compass work but more primitive means as well, such as celestial navigation, contouring, and terrain association—is paramount to the success of a guerrilla unit operating in a rural environment, regardless of whether that environment is an alpine forest, a jungle, a desert, or the taiga (the wet, coniferous forest below the tundra). Every single man must be able to go it alone anywhere, with or without map and compass. This is important for the conduct of patrols, wilderness survival, and escape and evasion. The guerrilla who lacks such skills is a threat to the success of the unit.

Fieldcraft

Mans called this "junglecraft" because that's where he was fighting the CTs, but since you may find yourself in places other than a jungle, we'll call it fieldcraft.

Lieutenant Colonel Mans quoted Field Marshal A.P. Wavell's famed perception of the skills a grunt must have: "He must be a combination of cat burglar, gunman, and poacher." This is a very accurate description of a guerrilla as well and works well here. (Field Marshal Erwin Rommel said of Wavell in *The Rommel Papers*, "The only one who showed a touch of brilliance was Wavell.") With regard to Wavell's description, Mans went on to say that "Certainly the good jungle warriors in Malaya needed all these antisocial characteristics and a surprisingly large number acquired them."

See Chapter 7 for additional guidance on fieldcraft.

Now that we know how to outthink the enemy, we can go underground.

CHAPTER 12

Going Underground

Guerrilla Tunnel Systems

"War is a singular art. I assure you that I have fought sixty battles, and I learned nothing but what I knew when I fought the first one."

—Napoleon Bonaparte to Gaspard Gourgard, 1815

Without question, the masters of guerrilla tunnel warfare were the Vietcong. Entire books have been written about the remarkable abilities of these savvy guerrillas when it came to building and utilizing secret tunnels and bunkers, and we can learn a great deal from them. One American general's dying wish, as he lay mortally wounded at the height of the Vietnam War after a sweep of a tunnel-infested area, was to meet the man who built those tunnels. Just meet him. It is this level of respect that shows us just how shrewd the Vietcong were.

The tunnel rat was born in Vietnam. A tunnel rat was a Marine or soldier, usually of slight stature, who would enter a tunnel with a pistol and a flashlight to try to flush out the Vietcong hiding therein. It was extremely dangerous work, and soon it became clear that the secret tunnels and associated subterranean bunkers were marvels of clever engineering and audacity. One sergeant major related to me an incident he witnessed with his unit in the Central Highlands where a Soviet-built T-55 tank belonging to the NVA was seen coming down a trail. It turned right and entered the jungle.

And disappeared.

Within minutes Marines were swarming in the jungle where the tank had just been seen, but the huge metallic beast was gone; it had disappeared underground. Despite hours of frantic searching, the tank was never seen again. Aircraft were called in to bomb the area with the hope of the bombs collapsing the concealed bunker the tank had pulled into, but to no avail.

But tunnels are not without dangers to the guerrillas. A very useful tactic that Marines and selected Army units used in South Vietnam in areas where tunnels were a problem was to leave a two-man sniper team behind to watch the area after the main body departed. According to 1st Marine Division records, in one eight-month period Marine snipers chalked up nearly 500 confirmed kills using this technique, which exploits the fact that the guerrillas must come out of their tunnels and bunkers sometime. And poor construction techniques and mistakes made by the guerrillas while they are in the tunnel/bunker complex can lead to the system's discovery by counterguerrilla forces as well. For instance, tunnels dug without sufficient turns and sumps to prevent smoke from filtering through the entire system will be vulnerable to being flooded with smoke as a method of determining where the other entrances to the tunnel are. Tunnels that weren't built so that one entrance can be covered by fire from another entrance are also vulnerable, as are tunnels that contain no booby traps, false walls, trapdoors, and hidden exits/entrances. The point here is that if you are going to build a tunnel/bunker complex, build it right or not at all.

TUNNEL PROS

The advantages of tunnel and subterranean bunker usage are many. They can be used as follows:

- for evasion when being pursued
- for staging food, weapons, ammunition, and equipment well forward in enemy areas
- as hospitals, rest and recreation facilities, observation posts, sniper positions, and infiltration routes
- for conducting briefings and training
- as repair facilities for weapons and equipment
- as command and control facilities
- as morale builders for harried guerrillas

TUNNEL CONS

Attentive counterguerrilla forces are adept at finding and destroying tunnels and underground bunkers. They look for the following:

- things out of place, such as small food plots hidden in undergrowth away from habitation
- wisps of smoke with no apparent source (from cooking fires underground; the odor of food cooking is another giveaway if that smell isn't near a house or hut)
- the odor of diesel fumes (from an underground generator)
- small piles of dirt or spoil spread out on the ground with no other explanation for its being there
- guerrillas all heading in one direction after being detected with no apparent avenue of escape
- continual sniper fire or harassing fire from one area that has been swept by conventional troops but with negative results
- ventilation shafts (usually hidden in clumps of brush or under logs or rocks)
- false canopies (extra brush and branches placed in living vegetation above a tunnel)
- entrance
- the smell or sight of human feces that appears out of place
- a faint trail leading into a thicket and then disappearing
- a cluster of scuff marks, footprints, and hand-holds on nearby trees or bushes with no obvious explanation

- an individual that appears to be by himself and has no obvious reason for being where he is

Counterguerrilla forces operating against guerrilla units known to use tunnel systems are trained to look for entrances in certain areas. They focus on buildings—where tunnel entrances can be hidden under sections of false flooring and under cooking pots hanging over fires on the ground (usually in a corner), and even beneath food stock piles—and anywhere that would allow the guerrilla better observation of the surrounding area: just off trails in a thicket, beside streams, and in hedgerows. The guerrillas must be sure to avoid patterning themselves in this manner. Imagination and ingenuity are the hallmarks of a good tunnel system. Sometimes the most obvious place is a good location for an entrance, and sometimes a good place is ingenious. An obvious place could be a secret entrance built halfway down the town well (make sure the actual entrance looks just like the wall of the well) or in the base of a bomb crater. If there are burned out vehicles lying around that are being taken for granted by the soldiers as just part of the scenery of war, a tunnel entrance could be run into the underside of a vehicle with part of the wreckage serving as the door cover.

To further avoid detection, fill in the initial excavation shaft that is dug to get to the level on which the main tunnel complex will be located below. This way, should someone who helped dig the excavation shaft be captured and interrogated, he won't be able to tell the enemy where the old entrance is because it will no longer be there. It is also wise to keep the system's whereabouts a secret from civilians so that, if questioned, they won't be able to divulge where it is either.

COUNTERING THE TUNNEL SWEEP-AND-DESTROY OPERATION

Modern counterguerrilla forces conduct tunnel sweeps with a force whose size is determined by the size of the area they are to search. Seldom is this force smaller than a platoon (anywhere between 25 and 40 troops), and it is usually no larger than a company (100 to about 160 troops). If the discovery of a major system is the goal, occasionally the enemy will bring in a battalion, but this isn't often the case. In any case, the unit is broken down into squads, which are given sectors to search.

Unit Task Organization

The enemy will divide into three primary groups—the search unit, security unit, and reserve unit.

Search Unit

These are the troops that will be doing the actual looking around for a hole leading to the tunnel/bunker system. They will try to get into every imaginable place to find a hole and will be looking hard for anything that might indicate a possible entrance. This team might also have its own security detail to supplement the perimeter security unit. Search teams that are effective are methodical and patient, and they don't care what lengths they have to go to in order to find a tunnel.

Security Unit

Securing the perimeter of the search area will be the security unit. Its job is to prevent attacks from outside the area and be watchful for guerrillas trying to sneak out. The security unit can employ observation posts, automatic weapons positions with cleared fields of interlocking fire, and sniper teams. The most modern counterguerrilla units may also employ remotely piloted vehicles to watch the area from above with a real-time video camera.

Reserve Unit

These are the troops who back up the security team and assist in keeping the area cordoned off to unauthorized foot and vehicle traffic. The reserve unit also contains the headquarters element. (The headquarters element contains the company or platoon commander, but it is unlikely that he will stay right with the headquarters element and sit on his duff. Most company-grade officers will go from subordinate unit to subordinate unit—a company commander will go from platoon to platoon, a platoon commander from squad to squad—to personally monitor their progress. A sniper should watch for this man and his ever-present radio operator. It is easy to identify this pair moving from unit to unit, with one of them always handing the radio handset to the other. If the tactical situation permits, the guerrilla sniper should kill this officer.)

Tunnel Infiltration and Destruction Techniques

The following will dictate the infiltration and destruction techniques employed by the enemy:

- the level of training they have received
- their fiscal resources (what they can afford)
- leadership ability (including that of any foreign advisors they may have with them, who may consist of experienced soldiers—officer and enlisted—or more clandestine types from the military branch of the supporting nation's intelligence service)

By knowing his enemy, the guerrilla can often predict what techniques and equipment they will use; therefore, the guerrilla can work to nullify both.

Demolitions, Grenades, and Bombs and Artillery

The use of demolitions (usually satchel charges) and grenades is commonplace when destroying tunnels. Given this, the guerrilla can construct tunnel entrances that thwart their use.

The illustration at right shows a typical tunnel entrance with a gas sump and a demolition sump. When intending to collapse the tunnel with demolitions without first sending in tunnel rats, a soldier will usually clear the entrance with either a burst of

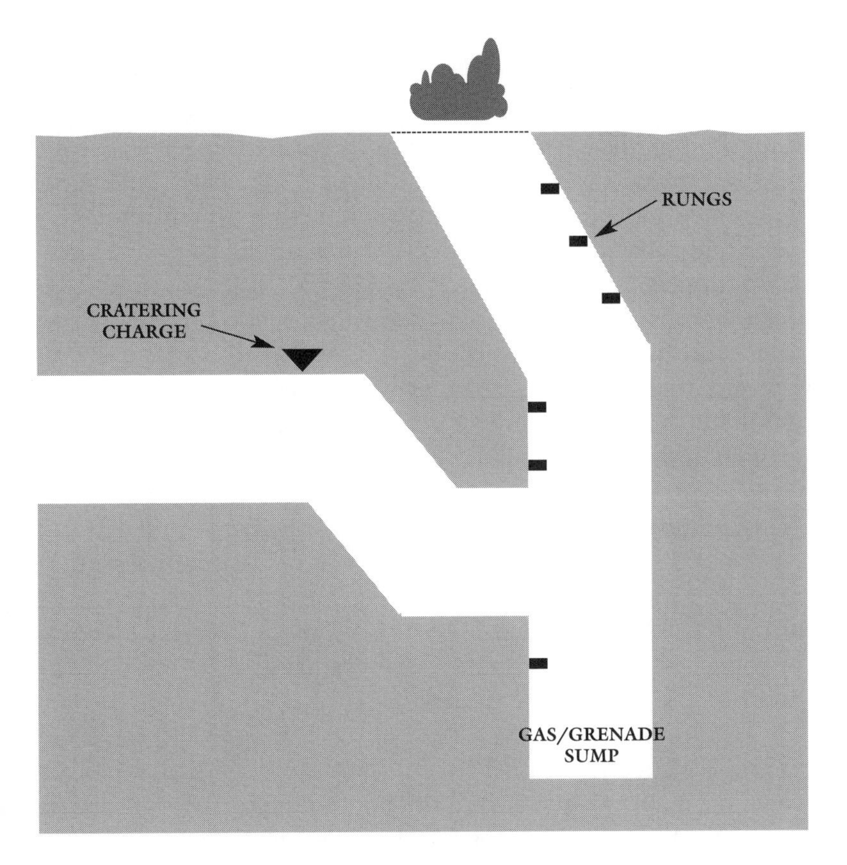

A TYPICAL TUNNEL ENTRANCE WITH A COMBINATION GAS/GRENADE (DEMO) SUMP

automatic weapons fire or a grenade and then enter the tunnel with a satchel charge. He will probably suspect a gas sump near the entrance and will go by it to try to find the main vertical shaft. Unless he is very brave, he will likely drop the charge down the main vertical shaft and retreat, then detonate the charge once he is clear of the entrance. A demo sump at the base of the vertical shaft will catch the charge and allow it to explode harmlessly. Above the gas sump on the second horizontal stretch, the guerrilla can place a hidden charge that he will detonate after the enemy's charge has gone off in the sump. This small charge is just strong enough to collapse this section of the tunnel so that, should the enemy reenter the tunnel to see if their charge worked, they will find the tunnel collapsed just inside and hopefully think it did. When the enemy depart, the guerrillas dig a new entrance; never re-excavate the old one, since the enemy will remember where it was and check it from time to time.

Bombs and artillery with delay fuses might be used to collapse tunnels. Artillery is much less effective than powerful fuse-delay bombs. When the enemy has heavy air power, tunnel systems must be very deep and heavily reinforced. Entrances dug beneath thick canopy are somewhat protected as they help to detonate the bombs and artillery shells before they strike the ground and have a chance to burrow underground.

Gas and Smoke

Smoke, tear gas, and white or red phosphorous from grenades, and, in some cases, special generators that force smoke into a tunnel can be used. These methods can be very effective provided the guerrillas do not have gas masks. Therefore, every guerrilla should have a gas mask.

Besides being used as a flushing tool designed to force the guerrilla to flee the tunnel system, smoke can be used to locate ventilation shafts—the smoke finds its way into the shaft, drifts upward, and is then seen by watchful soldiers. When the enemy can see smoke rising from several ventilation shafts, they get a rough idea as to the layout of the system below their feet—knowledge they can easily use to help destroy the tunnel.

The enemy may elect to roll fragmentation grenades down these ventilation shafts. This means that the guerrilla must build a grenade sump into the shaft so that no rolling grenade makes it to the main tunnel or bunker. Most fragmentation grenades have a 3- to 5-second fuse, so the sump should be no more than a meter or so from the surface. A sump so placed will catch the grenade before it has a chance to roll very far down the shaft, but far enough so that the soldier who rolled it thinks it did the trick.

Ventilation holes that are more narrow than a standard grenade may not be useful because of their reduced ability to

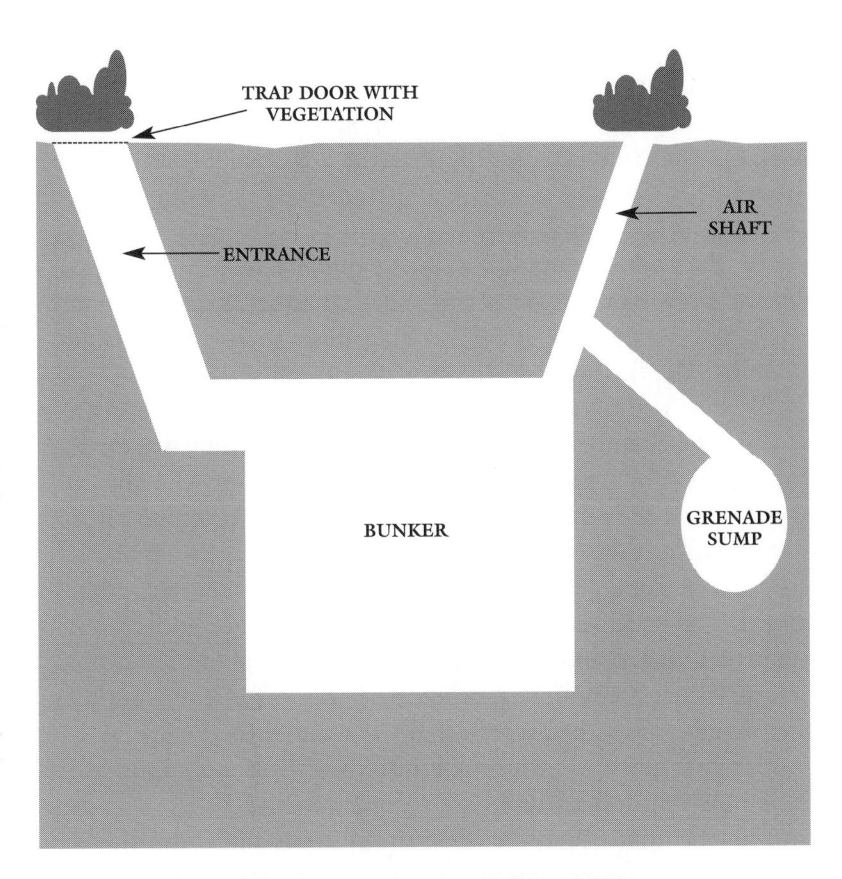

TUNNEL AND BUNKER SYSTEM WITH GRENADE SUMP AND AIR SHAFT

move air, and when dealing with advanced intelligence agency advisors the guerrillas may find that those advisors carry grenades that are half the size of a standard grenade.

Mine Detector

Mines made of metal should not be used near tunnel complexes because of the ease with which they can be detected by a mine detector; however, plastic mines (which are commonplace today) might be used. Better than mines are booby traps, which can be rigged above ground in many ways.

Any type of metal (mine) detector can locate metal below ground, including ammunition, weapons, comm gear, and so on stored in the complex. But the metal detector is only as good as the soldier operating it, and being human, that soldier is often easily led astray. Here's how.

As the tunnel complex is being built, a team of guerrillas should be collecting as many shell fragments from enemy artillery and mortars as possible. They place these in the ground at various depths between the tunnel and the surface.

When the detector indicates metal below, the enemy operator steps back and allows another soldier to carefully dig for the "mine," but he only finds metal fragments from what he will assume must have been a previous arty or mortar strike on the area. When the guerrillas have placed several fragments at each spot above storage areas holding metal, oftentimes, the enemy soldiers will get tired of detecting metal and only finding fragments, and they will start to become lackadaisical in their duties, believing that every time the detector goes off it is just another shell fragment and not worth their time and effort.

Another trick, a little more insidious, is to rig mines with antitampering devices such as a mercury switch. When a soldier touches the mine, the mercury in the vial rolls and completes the connection, thus detonating the mine. This is usually only good for one event, however, because every time after that the soldiers will just blow the mine in place rather than trying to remove it.

Dogs

The use of dogs is quite rare when clearing a tunnel, but it may be common in some units for locating tunnels. Dogs with good noses like Labradors, golden retrievers, German shepherds, and beagles can be trained to detect a great many things, tunnel entrances among them, but most dogs don't like going into narrow, dark tunnels and they aren't that effective when down there because of booby traps and difficult passages, such as vertical shafts.

It may be useful to engage the handler-dog team. Shoot the handler, not the dog. Why? This may sound strange, but shooting a man's dog is likely to cause more anger within the enemy unit than shooting the handler. Humans often form the strongest of emotional bonds to dogs, and they get highly upset when a dog is shot. This could easily result in a substantially greater level of determination and revenge being demonstrated by the soldiers. Also, dog handlers receive just as much training as the dog, and the two are a bonded team. It might take longer to replace the handler than the dog.

Night Vision Goggles and Flashlights

Digging tunnels with frequent turns can lessen the effectiveness of devices such as these; even the best night vision goggles (NVGs) can't see around corners.

If electricity is available within the tunnel system and the rat is wearing older model NVGs that don't have the ability to counter sudden bright lights quickly, rig up a lightbulb right in the middle of the tunnel. Keep it off until the guerrilla is right in front of it and then turn it on. The sudden bright light will "white out" his goggles and temporarily blind him. Then do what you have to do.

Weapons

Tunnel rats carry pistols. Like NVGs and flashlights, bullets don't go around corners (at least not the ones I am familiar with, although some acquaintances of mine—don't even ask; they even scare me—in a certain government agency say their organization is working on it). Frequent turns help reduce the effectiveness of pistols.

Flame weapons come in four types: those that fire an incendiary projectile (such as the old M202 Multi-shot Incendiary Rocket Launcher, white or red phosphorous shells, or Fuel Air Explosive bombs [FAE]), those that fire open flames (the classic flamethrower), those that burn after being detonated (certain mines and booby traps), and those that burn after being poured (gasoline). Only poured flammable liquids are effective in tunnels in most situations, and even then modern counterguerrilla forces seldom use them because they are so easily foiled through the use of sumps.

Some Evil Tricks

Besides avoiding detection in the first place, one of the guerrilla's goals in tunnel warfare is to discourage the enemy from entering the tunnel. This is done by making tunnels extremely dangerous to enter and clear. All tunnel tricks must be subtly marked on both sides (coming from both directions) in such a way that every guerrilla will know not only that there is a trick there but what kind of trick it is and therefore not fall prey to it. Use whatever system works best for you, but make sure every guerrilla knows that system by heart. Ensure that the mark won't be readily noticed by a tunnel rat and identified as a warning. Simple marks like a single bullet left lying in the tunnel might make a rat think that a guerrilla just dropped it there unknowingly, but it is really there to tell the guerrillas that a certain trick is just ahead.

Miner's Nightmare

More than anything, a miner fears a cave-in. All tunnel rats fear cave-ins just as much as the miner does.

Walls and ceilings of tunnels can be weakened and rigged to collapse through the use of remotely detonated demolitions carefully placed behind opposing walls and in the ceiling above those walls. The charges can be detonated by command or by the rat tripping them with a hidden pressure switch in the floor. A modification of this is to have the pressure switch eight or nine feet past the charge. This is good to terrorize the rat by trapping him inside the tunnel, or for when two rats are in the tunnel, one right behind the other. The second rat is buried alive by the

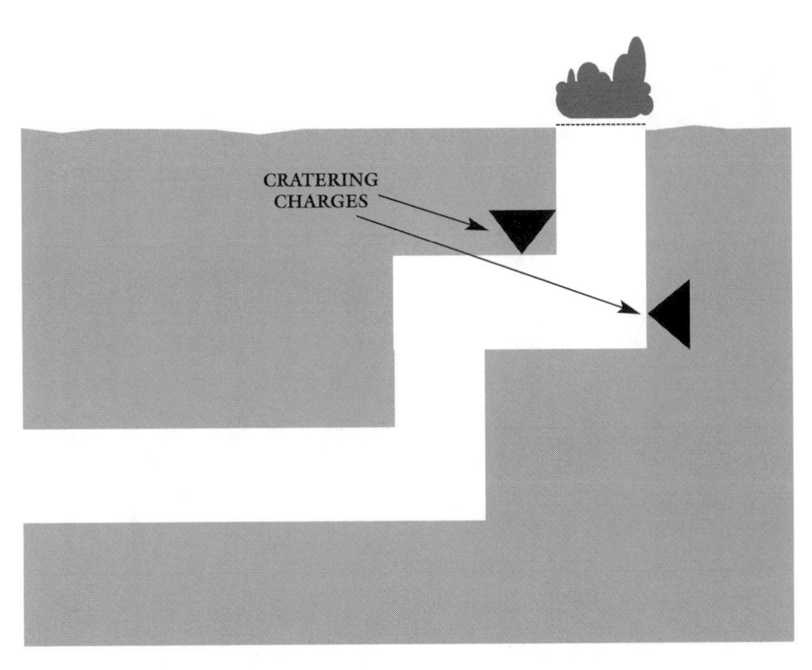

TUNNEL CAVE-IN CRATERING CHARGE

actions of the first rat, who tripped the device, and who then has to either try to dig his partner out or find another exit. If the first man lives to tell the tale, he now has to continue fighting knowing that he caused his pal's death. Tunnel warfare is always a game of psychological warfare.

At a minimum, a tunnel collapse with a soldier inside slows the search down as would-be search teams come to help try to excavate the trapped soldier. If a less than aggressive commander is in charge he may even call off the search and withdraw.

False Floor Pungy Pit

Taking yet another cue from the Vietcong, dig a pungy pit in the tunnel floor and cover it with a false floor so that the tunnel rat's hand goes through the false floor as he crawls forward. Rare is the tunnel rat who will continue the search once his hand has been perforated by several sharpened stakes.

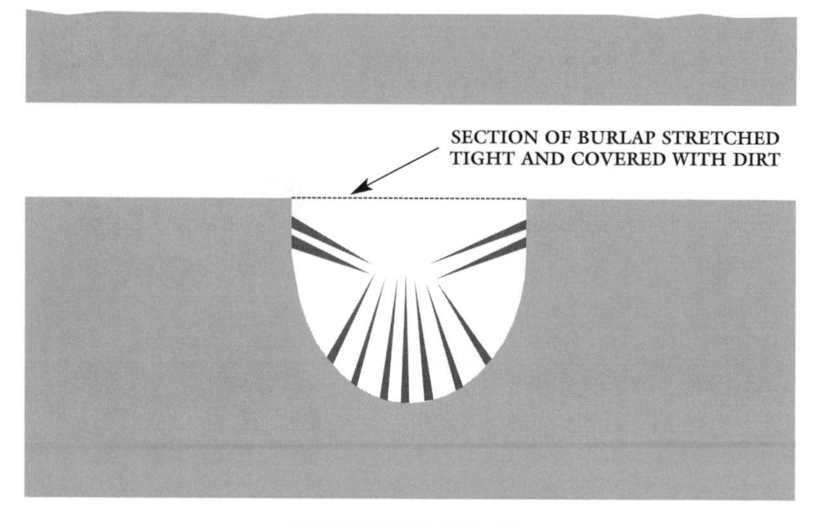

TUNNEL PUNGY PIT

A modification of this is to place a very aggressive, highly venomous snake in the pit along with the pungy stakes. Now the rat has holes in his hand from stakes and a nasty snakebite; he will depart the tunnel immediately to seek treatment, and the other rats won't be especially keen on going into the tunnel Corporal Deadguy just came screaming out of.

Corner Shelf

About head height for a crawling man and immediately around a 90-degree corner in the tunnel, dig a shelf into the wall. Place a deadly snake on it and tie its tail to a stake with a strong piece of string (have another man hold the snake behind the head to prevent your being bitten as you do this). As the guerrilla turns the corner, the idea is for the snake to strike. A head wound inflicted by a snake like a bushmaster, western or eastern diamondback rattlesnake, krait, fer-de-lance, puff adder, or other equally dangerous reptile is more likely to kill the soldier than a wound elsewhere on the body.

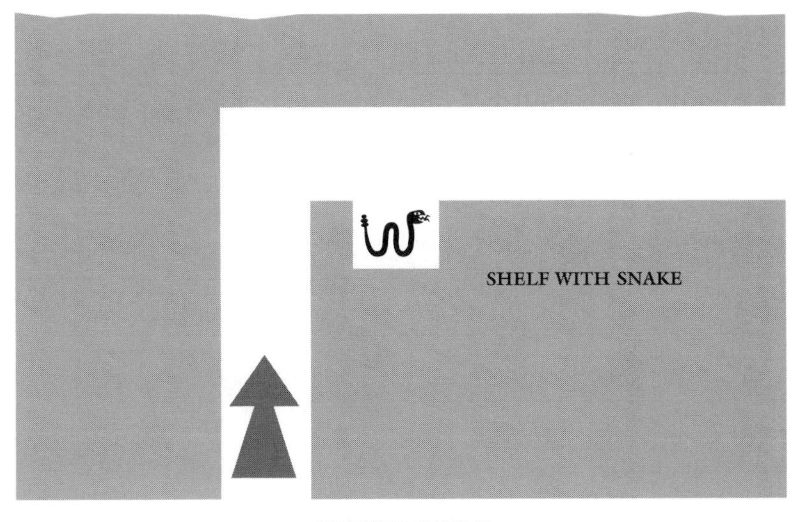

CORNER SHELF

Take a Shave

Single-edged razor blades can be placed in the floor in bunches about an inch apart and barely covered with dirt. The rat places his hand on them and is cut to shreds. As you can do with pungy stakes, apply feces to the blades to increase the chance of infection. If you are operating in areas with certain species of wildlife that are highly toxic, such as the poison arrow frog of the South American rainforest (a colorful little frog that is unbelievably toxic just to the touch; Jivaro Indians use leaves to pick the little buggers up, lest they be poisoned themselves, and then use the poison the frog's skin releases on their hunting arrows and blowgun darts), even more evil can be done.

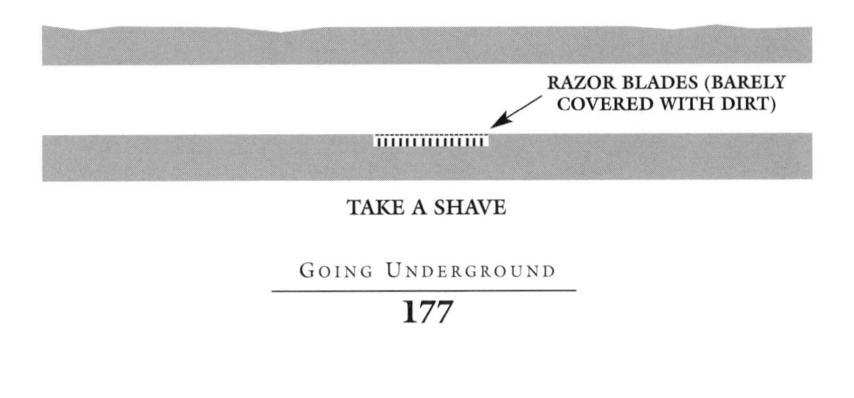

TAKE A SHAVE

Dump Site

Although not fatal, a pile of human feces placed in a shallow hole in the floor and then barely covered with dirt will prove highly annoying and disgusting to the tunnel rat.

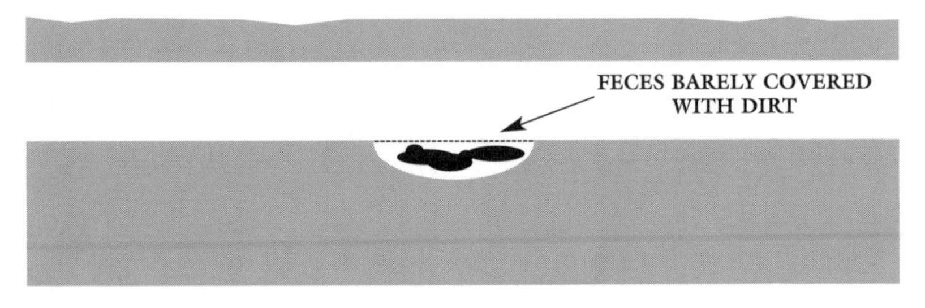

DUMP SITE

Shell Game

Rig a detectable but not blatantly obvious booby trap in the tunnel. This one isn't intended to hurt the rat, but the unseen antipersonnel mine or booby trap 10 feet before it is. Most tunnel rats, when they see a booby trap, will crawl forward to check it out and see if it can be bypassed or rendered inert. They often forget to look for better-hidden devices between the point they detect the ruse booby trap and the booby trap itself.

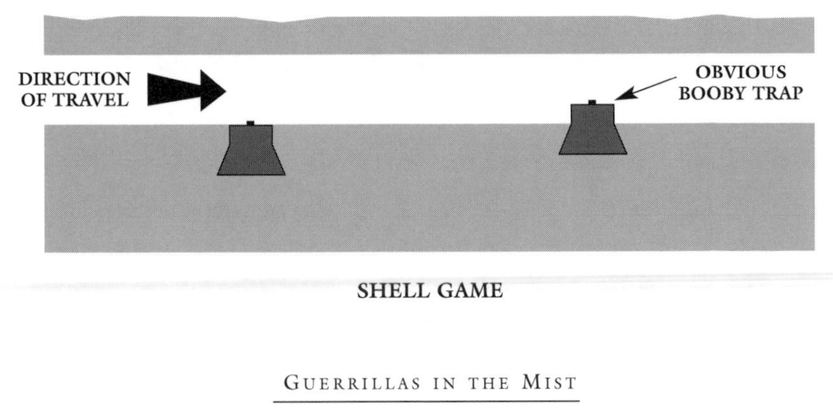

SHELL GAME

Poker

At chest height for a crawling man, a hole is drilled through the tunnel wall into another chamber the rat doesn't know is there. In that chamber is a guerrilla with a spear. The spear's tip and the hole the spear is in are concealed from the crawling soldier by a small piece of false wall that is thin enough for the spear to pierce easily. From another secret hole unseen by the rat, the guerrilla watches. When the rat gets in front of the false section of wall, the guerrilla rams the spear forward and skewers the tunnel rat. (A firearm can be used in place of the spear.)

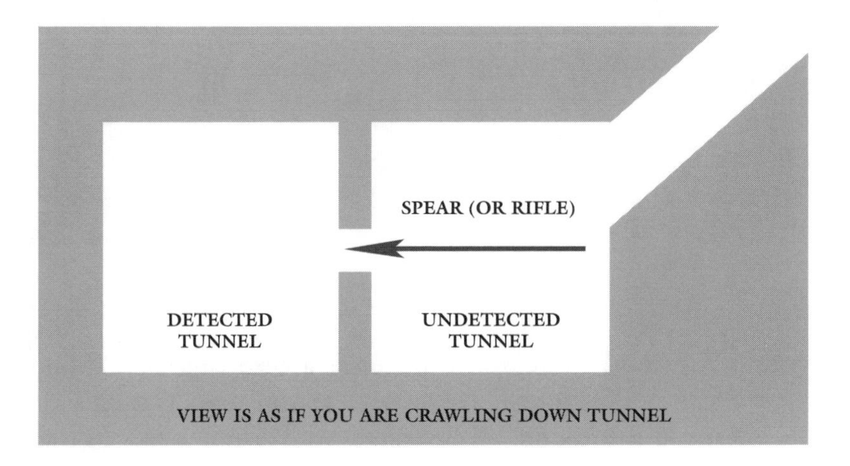

POKER

Ben Franklin

Run a metal bar about a foot out from the tunnel wall so that it appears to be just a piece of rebar reinforcing, or run it out of a wall forming a vertical shaft to appear as a rung for climbing in the shaft. Now attach a live wire with a few amps running through it to the other end of the bar. When the tunnel rat grabs the bar, he's toast.

These are only some of the tunnel rat deterrents a guerrilla force can employ. Use your imagination to hurt the enemy, and

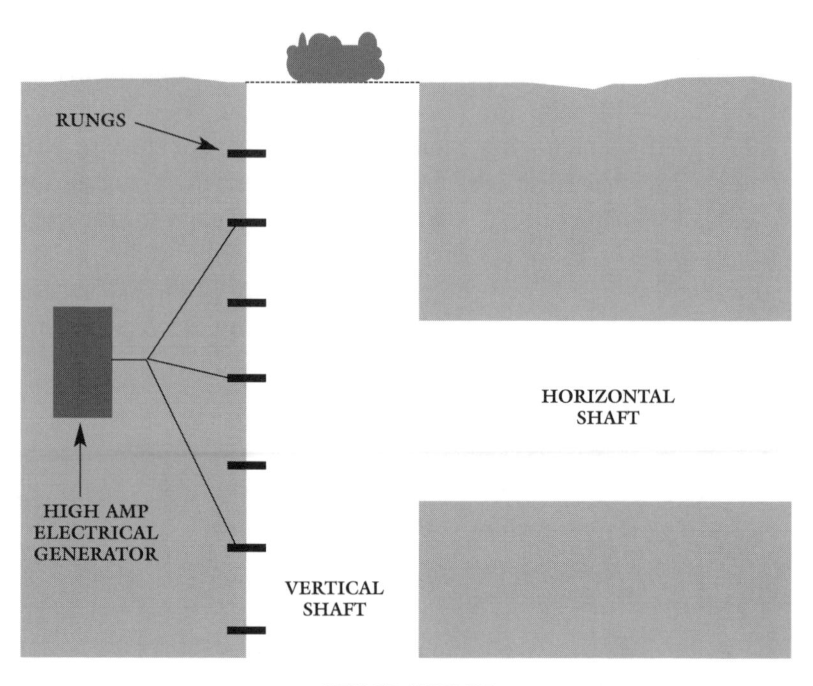

BEN FRANKLIN

don't forget to make the experience a terrifying one. Mind games are terrible for the tunnel rat. Try this one: place a soda can right in the middle of the tunnel with a note attached that reads, "No, this isn't a booby trap, but there are some ahead. Have a nice day. It will be your last with eyesight." Or, "Poisonous snake ahead." Remember Madonna's line in her song "Vogue": "All you need is your own imagination, so use it, that's what it's for."

TUNNEL CLEARANCE AND DESTRUCTION SOPS

Most counterguerrilla units will have set SOPs for clearing and destroying tunnels. Because of this set-in-their-ways approach, the guerrilla can observe the enemy clearing and destroying tunnels and make plans to counter those techniques. Modern counterguerrilla forces often use a four-step procedure.

Burst of Fire

To kill any enemy right near the entrance or force them to move deeper into the tunnel, a soldier will fire a long burst of automatic weapons fire directly into the tunnel opening straight through the hole's door. (If the door is thin wood or made of thatched grass or branches, a flamethrower—a weapon that the Marines used very effectively during World War II to flush Japs from tunnels and "spider" holes—can be used for this as well.)

The guerrilla can reduce the effectiveness of this first step by building two turns into the tunnel immediately after the entrance hole.

This Japanese soldier found out the hard way that a Marine with a flamethrower can be an effective countertunnel system. (Department of Defense photo.)

Even the darkness of the tunnel's entrance sends shivers down the tunnel rat's spine. What waits inside?

Door Destruction

If the door is still intact, a demo charge or grenade is placed on it to clear it and also detonate mines and booby traps that may be associated with the door.

There isn't a lot the guerrilla can do about this.

Drop Charges and Grenades

The third step is to drop a demo charge or fragmentation grenade down the tunnel itself.

As already discussed, sumps and turns help reduce the effectiveness of this step. The enemy may choose to use gas or smoke grenades at this point, too, and the countermeasures for these have already been discussed as well.

Tunnel Rats

The final step comes with the introduction of tunnel rats into the system to clear any remaining guerrillas and gather documents, weapons, equipment, and so on, and then destroy the system from within.

All the possible methods for countering the rats were listed earlier.

TUNNEL DESTRUCTION

In most cases, the enemy will choose to destroy the tunnel complex with explosives. Therefore, it is wise to construct tunnels with false walls and trapdoors hiding reinforced bunkers and connecting tunnels. The enemy, when using explosives, might use one or more techniques.

Bangalore Torpedoes

Most often used to breach perimeter defenses, a series of bangalore torpedoes placed throughout the complex and detonated simultaneously can bring down the entire complex. Reinforcing walls and ceilings may have little effect. It might be feasible for a sniper to target the bangalores if the enemy is foolish enough to stack them all in one place in preparation for use.

Cratering Charges

These can be very effective against subterranean bunkers. The deeper the bunker and heavier the reinforcement, the better for the guerrillas.

Block Charges

The primary benefit of a block charge is that it can be tamped right against the tunnel ceiling and, when detonated, could have a ripple effect that collapses the whole tunnel. However, this usually takes a charge of at least 10 pounds.

Satchel Charges

These can be placed or tossed into shallow tunnels and bunkers and can be effective in collapsing them, but the destruction does not often go far beyond the room or tunnel section in which the charge exploded.

Shaped Charges

Shaped charges can be set deep within a system and detonated either upward or downward to achieve a lot of destruction. Reinforcing walls doesn't often do much good against such a charge.

TUNNEL AND BUNKER COMPLEX CONSTRUCTION

Tunnel and bunker systems run from the very simple to the very intricate. What gets built where depends entirely on the tactical situation and available manpower and logistics.

Simple Holes and Hides

These are the easiest to construct and take minimal manpower. They can be built in enemy areas as infiltration supporters and for when the guerrillas must break contact and hide quickly. Normally they are only large enough for a maximum of three guerrillas, and that is often pushing it, with one- and two-man holes being more common. They can be placed below huts or under clumps of vegetation, and they can even have underwater entrances that are accessed from drainage ditches and streams. Two air holes are the norm, but in some cases one will suffice.

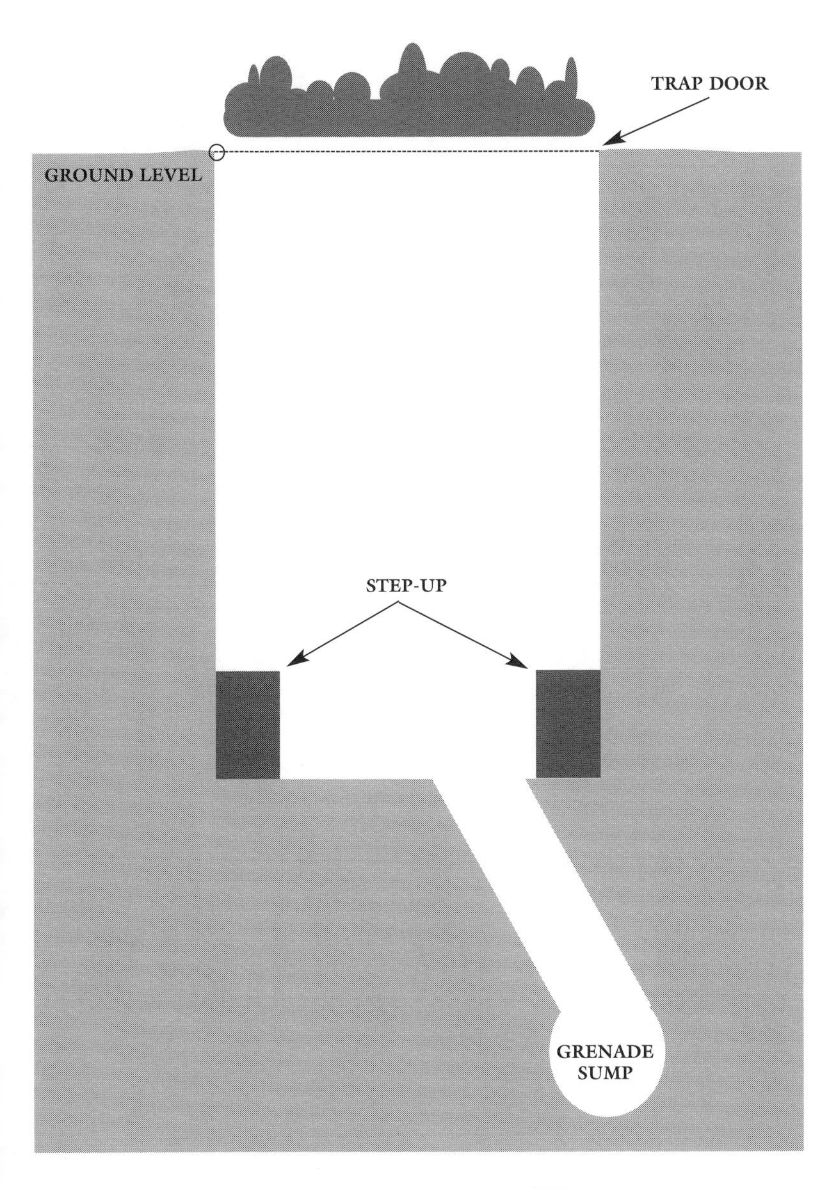

SIMPLE FIGHTING HOLE/HIDE

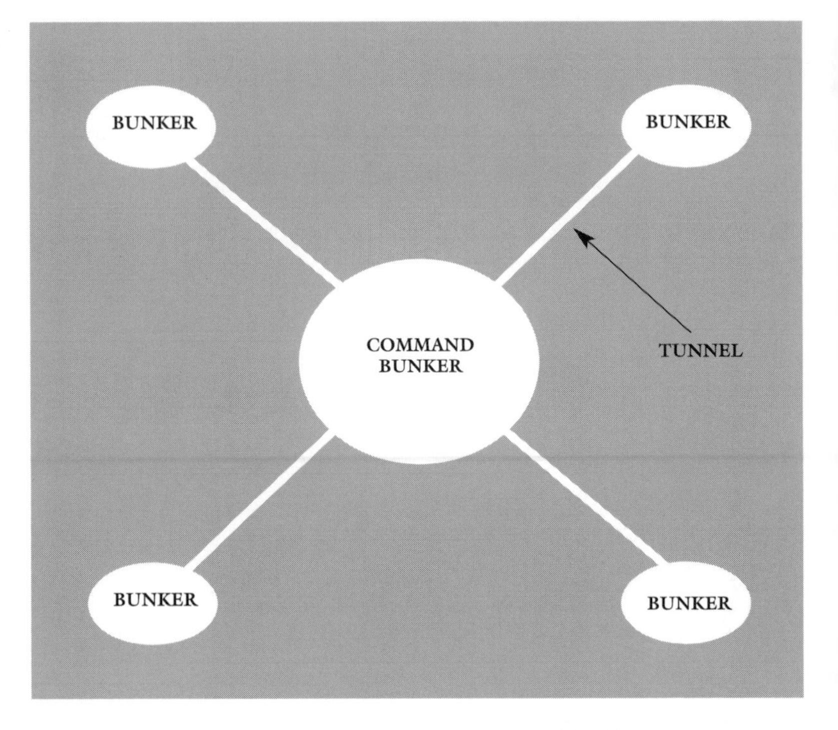

UNFORTIFIED UNDERGROUND BASE CAMP

Unfortified Base Camps

Since they are unfortified, these base camps are maintained in regions controlled by the guerrillas and are used for logistical and command and control sites. Because they are in guerrilla-controlled areas they are often very complex and deep. A variety of rooms and room sizes can be used. Trenches and supporting fighting positions are not used.

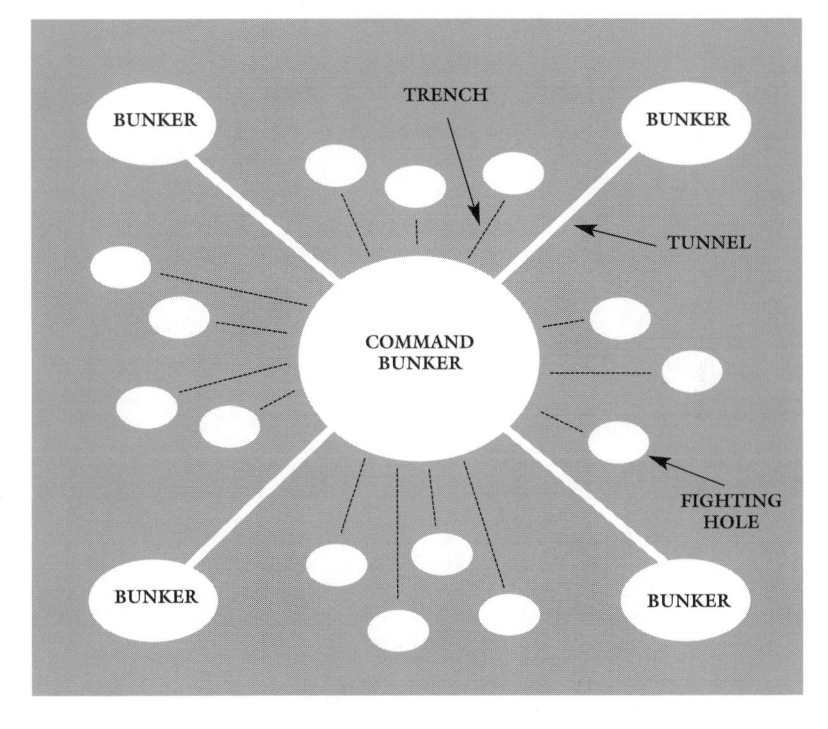

FORTIFIED UNDERGROUND BASE CAMP

Fortified Base Camps

The most complex of systems, the fortified base camp is centered around a command bunker that often protrudes up to two feet above the surface with firing and observation ports. These camps are in enemy areas. Radiating out for long distances (often hundreds of yards) from the command bunker are tunnels that lead to outlying bunkers of similar design to the command bunker, and these outlying bunkers all have mutually supporting fields of fire.

In between the tunnels are trenches leading to individual fighting positions at various distances from the command bunker. These, too, offer mutual support.

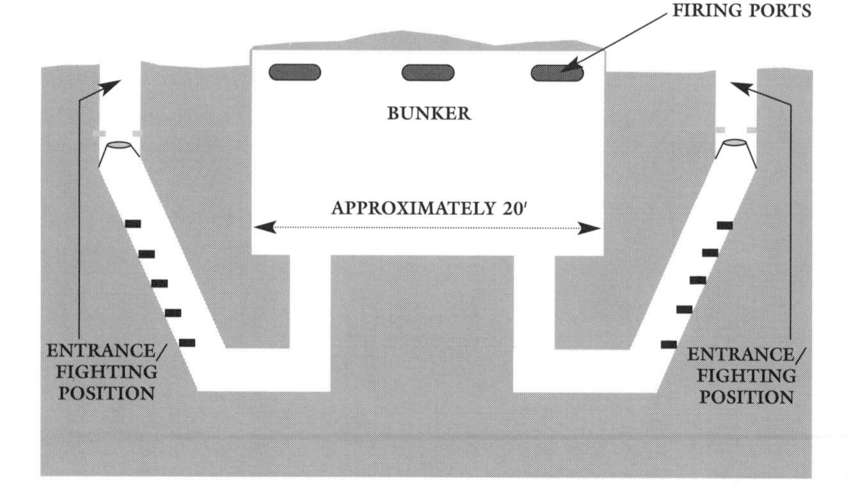

BUNKER WITH TUNNELS TO COMBINATION ENTRANCE/FIGHTING POSITIONS

Bunkers

Bunkers can be made of assorted materials ranging from bricks, cinder blocks, and concrete to logs and sheet metal and any combination of these. Some have firing ports, others don't, but all are protected from overhead fire and have tunnels leading to and from them.

Guerrillas must be experts in designing, building, maintaining, and operating from bunkers. They are an excellent combat multiplier for a force that is intrinsically weaker than its enemy. The guerrilla force that uses them well stands an improved chance of victory.

CHAPTER 13

Into the Streets

Guerrilla Tactics in Built-Up Areas

"My solution to the problem would be to tell [the Vietcong] they've got to draw in their horns and stop their aggression or we're going to bomb them back into the Stone Age."

—Gen. Curtis E. LeMay, 1965

Guerrilla movements are more and more likely to be at least partly involved in built-up areas such as the inner city, industrial areas, and the suburbs. This is true because our world populations continue to grow and there appears to be no chance of their stabilizing, much less dropping.

In the near past, guerrilla wars have been fought in cities like London, Derry, and Belfast (although the IRA has made the terrible and unforgivable mistake of turning to terrorism rather than abiding by an honorable and genuine guerrilla war), Kabul, Madrid (where the Basque separatists, known as ETA, also made the mistake of going the terrorist route), Colombo, Grozny, Saigon, and Lima (again, the latter suffering terrorism at the hands of the Shining Path and Tupac Amaru) to name a few. Future guerrilla wars will increase in the cities and their surrounding communities.

In an urban or suburban environment, the guerrillas will be unable (and unwise) to muster and maintain large standing forces. Instead, individuals and small cells of guerrillas will be operating in such a way that only one member of each cell will know a member of another cell. This increases internal security. The individuals and cells will focus on the following:

- ambushing vehicular and foot patrols
- bombing enemy supply points and other logistical or command and control facilities
- sabotaging public services such as mass transit, communications, sewer systems, garbage collection systems, medical services, power plants, port facilities, and security services
- attacking industrial targets that directly support the government and city
- sniping enemy patrols, roadblocks, check points, and leaders
- conducting bomb or rocket attacks on government buildings
- provoking the enemy forces in such a way that they lose control and injure or kill civilians

The guerrillas must make all of these actions appear to be the fault of the government. They must make a great effort to prevent unnecessary civilian injuries and deaths that will be seen as being the fault of the guerrillas. When the people believe that the guerrillas are on their side and the government is unable or unwilling to fight them, the battle is nearly won.

Modern guerrilla armies must plan for and conduct operations in the following types of built-up areas:

- urban housing areas, including tenements and apartments
- suburbs, including free-standing houses, condominiums, townhouses, and apartments
- industrial areas
- clusters of various buildings in a variety of settings (associated with roads/highways, railroads, etc.)

The wise guerrilla leader will never assume that he will be operating only in one type of area, for as the tactical and strategic situations change, the more likely it is that the guerrillas will find themselves having to change along with them. This means you may have to conduct a demolition raid on a major port oil storage facility in a city of more than half a million inhabitants one night and stage a convoy ambush along a suburban side street the next.

TACTICAL FACTORS

Both the guerrilla and the enemy must take into account several factors involving their respective operations in built-up areas. Each factor can either be a hindrance or an asset to the parties, depending on how they are used. In the built-up arena, innovation plays a special and often incisive role in determining the outcome of a battle.

The Surrounding Terrain

Conventional armies fighting in a guerrilla war tend to think statically and are therefore largely predictable; their field manuals and institutional training dogma make this so. Better yet, if the enemy do not enjoy a decentralized philosophy of command and the guerrillas do, they expose themselves to a much higher risk of destruction on the built-up area battlefield. There are five key terrain factors the guerrilla must consider when fighting in a built-up area.

Key Terrain

The enemy will be focusing on key terrain features within the built-up area. These include buildings that are capable of withstanding a lot of punishment and overlook likely avenues of approach, sewer and subway systems, bridges, railheads, port facilities, television and radio stations, tall buildings that could support plunging fire, and so on. By booby-trapping these places and registering indirect fire on them before the enemy gets there so that your fires will be immediately accurate, you increase your combat power.

Observation/Fields of Fire

Observation and fields of fire tend to be restricted in built-up areas because of the buildings, smoke, dust, and rubble. The enemy will therefore seek vantage points that are above the fracas in order to direct their fires. The guerrillas should use smoke, dust, and debris to conceal their movements and at the same time direct fire on enemy observation points, which should also be booby-trapped.

Obstacles

These are limited only by the guerrilla's imagination. Creative obstacles slow the enemy's progress and expose them to fire, and once exposed their actions become all the more pre-

Buildings can provide both cover and concealment if used properly.

dictable. Obstacles should channel the enemy into minefields and areas that are heavily booby-trapped, as well as areas where the guerrillas have open fields of fire to the enemy with automatic weapons and mortars. An excellent way to slow and channel the enemy's approach with no fire whatsoever is to make some realistic NATO chemical markers and place them along the enemy's axis of advance so that they have only one logical way to go. That way should be mined and covered by automatic weapons fire and mortars—hold your fire until the main body is in the kill zone, then lay waste to them.

Cover and Concealment

Cover provides protection from fire; concealment provides protection from detection. For the guerrilla, cover and/or concealment can, depending on the situation, take the following forms:

- vehicles
- buildings
- dumpsters
- crowds
- signs
- anything else that prevents the enemy from detecting or firing on you

Avenues of Approach

Avenues of approach include the following:

- highways/turnpikes
- railroads and subways
- waterways (no matter how small)
- roads and streets
- alleys
- sewer systems
- jogging/bicycle trails

One must never assume that a certain avenue of approach is unusable. Even the narrowest, most unlikely access route, such as a jogging trail through a park's woodlands or an old sewer system filled with rats, fetid water, and feces, can be used by a shrewd counterguerrilla commander to infiltrate the guerrillas' area of operations and catch them unaware at the worst possible moment. So intent are the U.S. Marines on effectively dealing with urban guerrillas that they have built elaborate MOUT (Military Operations on Urban Terrain) facilities that are realistic to the point of having sewer systems that Marines and aggressor units use to simulate urban guerrilla warfare; the miniature cities even have gas stations, stores, apartment buildings, streets, sidewalks, and parks.

The People

How the populace feels about the guerrillas will play a role—often a critical one—in the outcome of the guerrilla war. If the people do not feel that their lives are being made untenable by the government and the national or foreign forces supporting the government, the guerrilla movement will never succeed. The people must be shown that the government and its henchmen are the true enemy. The guerrillas must be adept at the following:

- stirring the people into action, such as strikes and antigovernment demonstrations
- arranging for the government to appear responsible for social woes such as racial tension and problems between various other groups

Remember that the people are the guerrillas' greatest asset if handled correctly.

The Enemy

How well the enemy are trained in counterguerrilla warfare,

the level of their leadership, and how committed they are to destroying the guerrillas all play a role in the outcome of the war. This is where the guerrillas' knowledge of the enemy comes into play, for without a clear and accurate understanding and assessment of the enemy, the guerrillas stand no chance of victory.

The Guerrillas

As with the enemy, the guerrillas must be committed, and they must have superior leadership, excellent training, and a reliable means of logistical support to be victorious. Nowhere in the realm of guerrilla warfare are security and small unit leadership more important than when fighting in built-up areas; it will usually be the NCOs who lead and conduct most urban guerrilla operations because of the smaller size of the individual guerrilla forces (cells). Here the decentralized philosophy of command shines its brightest.

Now for some night ops.

CHAPTER 14

Under Cover of Darkness

The Nocturnal Solution

"I am more afraid of our mistakes than our enemies' designs."

—Pericles to the Athenians, 432 B.C.

Guerrilla warfare is a game of outthinking the other guy. It is a game of striking at him in a bold, aggressive, and unexpected manner in a place and at a time that you believe give you an advantage from which you can win a decisive victory. To win at this game, the guerrilla must learn the enemy's tactics and in doing so learn how to prevent the enemy from using the tactics and weapons they believe will allow them to gain the upper hand and win. Therefore, guerrilla warfare is a game of preemption and timing, anticipation and guile, courage and knowledge.

The guerrilla force must train to conduct at least 70 percent of its operations under cover of darkness; the guerrillas must own the night to such a degree that the enemy greatly fear having to go after them after the sun sets. (It is interesting to note that one of the world's most respected counterguerrilla forces, the U.S. Marines, still conducts less than half of its training at night, this despite Marine doctrine indicating that at least 70 percent of Marine training should be nocturnal.) If the enemy soldier, when told he is to go on a night patrol to hunt a guerrilla unit, gets an immediate burning sensation in his guts and breaks out in a cold sweat upon learning of his next mission, then the guerrillas in that area are doing their job.

Darkness gives the following two important tactical advantages to the guerrilla force:

- It conceals the force so that it can remain undetected after setting in for an ambush.
- It allows the force to withdraw quickly once the ambush has been triggered and completed.

It can also allow the guerrillas to do the following:

- make up for what in the daytime would be a tactical disadvantage due to reduced combat power
- maintain contact with the enemy from a position of relative safety
- follow up on daytime successes

TACTICAL CONTROL MEASURES

The enemy are likely to use certain tactical control measures as they prepare for their night operations. The guerrilla force's job is to determine what and where those control measures are and strike a decisive gap in those measures before the enemy attack actually commences. By deceiving the enemy into thinking they have found a good target in such and such a location (which in reality is a ploy on the part of the guerrillas to get the enemy to come to a certain place), the guerrillas can often deduce much about how the enemy will get from their base camp to that target.

Assembly Area

Night attacks often call for an assembly area that is closer to the line of departure than one that would be used for a daylight attack. This means that the enemy are going to have to move out of their defensive position at night to get to that assembly area, and they will usually move in columns as they do so. But before they send out the first column they will deploy reconnaissance teams and scouts to reconnoiter the area they will be traveling through and the route they intend to use. The guerrillas must ensure that they remain undetected by these recon patrols and scouts, observing their movements and actions all the while. Also, the enemy may use aerial reconnaissance—manned or unmanned (remotely piloted drones)—to gather tactical information about the assembly area and the route planned to get there. Guerrillas must be extremely wary of aircraft and drones. Any reconnaissance measures used by the enemy can be helpful in determining the enemy's route of march.

Once the enemy recon and scout units have completed their mission, they can either return to the base camp and report their findings, or they may stay hidden near the assembly area and send their report back to the rear via radio. If they stay near the assembly area, the guerrillas must select an ambush site somewhere between the enemy's camp and the assembly area. This site must be far enough away from the reconnaissance teams and scouts to prevent their counterattacking the guerrillas, and it likewise must be situated far enough away from the enemy's main body's reserve force, which could also be deployed as a counterattack force. For additional insurance along these lines, the guerrillas may be able to set up a blocking or delaying force between these potential counterattack forces and the ambush site. Remember that a blocking force seeks to deny the enemy access to a certain area or avenue of approach, and a delaying force seeks to trade space (terrain) for time, with the idea of buying just enough time for the ambush force to withdraw.

Release Points

Release points are potential gaps because they are points where a higher commander relinquishes control of subordinate units to those units' commanders. If a release point is located at an easily recognized terrain feature and the guerrillas anticipate this, confusion can result in the enemy force when the ambush is sprung because of possible misunderstandings in command and control at that point. Unless the commander on the scene takes immediate and bold action to counter the ambush, the guerrillas stand a good chance of getting a decent body count. (But don't be fooled into thinking that body counts win wars; during the Vietnam War, America learned that this isn't always the case.)

Attack Positions

Attack positions are used only occasionally during night ops, and normally they are only occupied for brief periods of time due to their propensity for suffering spoiling attacks. However, the enemy force may need to pick up some special equipment or weapons along the way or receive additional instructions from higher command that require a quick stop for dissemination to the small unit leaders, and the attack position may be that place. Potential attack positions should be mined rather than ambushed, and the mines should be placed throughout the position.

Routes

If the enemy is lacking in reconnaissance measures, routes of march can be some of the easiest and most effective control measures along which to lay ambushes. These routes are usually chosen by terrain and their proximity to and attitude toward the enemy's objective. Oftentimes, guides are forward-deployed by the enemy maneuvering force to assist the smaller units in finding their lines of deployment. Guerrillas must be cautious to remain hidden from these guides.

If manpower permits, guerrilla scouts must be placed along all potential avenues of approach to watch for signs of enemy movement. Be aware of ruses; the enemy may send in recon and scout teams along routes they have no intention of using in order to draw guerrilla forces away from their true route.

Radios

The enemy will no doubt have radios from the squad level on up. Guerrilla sharpshooters should make priority targets those troops with radios on their backs and anyone who is called to speak into the handset by the radio operator—this is almost always a leader.

Markings

Sometimes the unit leader will mark his uniform in some way so that his men can easily tell where he is at all times, such as luminous tape ("cat eyes") in a particular pattern on the leader's helmet or hat. Again, sharpshooters should make anyone who looks different a priority target.

THE PLAN OF ATTACK

The enemy's plan of attack will consist of a scheme of maneuver and a fire support plan (indirect fire weapons that will support the enemy attack, such as mortars and artillery). How the guerrillas anticipate and deal with both of these elements will have a direct impact on the outcome of the ambush.

Scheme of Maneuver

A scheme of maneuver is a plan formulated that details the employment of all enemy units other than fire support. The scheme of maneuver includes all subordinate, attached, and supporting units (the latter of which aren't fire support but rather units used to support the main effort by acting as decoys).

Prior observed engagements and the enemy's own tactical and operational dogma can be invaluable in determining his scheme of maneuver. This is where the battle of wits between the guerrilla leader and the enemy leader begins. It's almost Abbott-Costelloian in concept, i.e., you are formulating your ambush plan on how you expect the enemy to maneuver—you know how he thinks. But the enemy knows you know how he thinks, so he might change his plan to counter that. Then again, you know that he knows you know how he thinks, so you go to Plan B in order to fool him. But he knows that you know that he knows that you know how he thinks, so . . . well, you see the problem.

The trick is to do something he doesn't expect and do so in such a way that it hurts him badly in a very short period of time—a matter of seconds—and then melt into the shadows so that he can't counterattack or pursue your withdrawing forces. This is the essence of an ambush.

I recall laying an ambush once with my six-man recon team. We had an M-60 machine gun along, and I laid in the ambush the way I thought would produce the best results. My machine gunner, Todd Ohman, said that my plan was okay but that he had a better one and that I should use his plan. I wanted to do it my way and didn't listen to Todd—when I should have—and the ambush produced only marginal results. Todd had anticipated a problem that did in fact present itself, but because I was in charge and still learning the trade at that point, I didn't listen. In fact, after the ambush was sprung, my team was very nearly ambushed by a counterambush team. Todd's plan would have avoided that team.

The next time, I listened to Todd.

Fire Support Plan

The enemy's fire support plan must be considered and planned against from the start. By utilizing real-time radio communications, where the ambush team has direct comm with the counterfire support team, the ambush commander can inform the counterfire support team that his ambush is in progress and that he should now prevent the enemy fire support unit from doing its job. So we see that this team is in fact the ambush team's fire support unit.

The enemy fire support unit's position must be reconnoitered carefully if it is to be engaged effectively; never assume that you can take it out easily with a single means. For instance, if you expect the enemy company commander to use his organic mortars as his primary means of fire support—because that's what he has done in the past and that's what his manuals say to do—and you intend to take those mortars out with sharpshooters, you would be dismayed to find that he has dug in his mortars and they are now very difficult or impossible to engage with direct-fire weapons like rifles.

If you first recon the mortar position you can better select a means of engagement. Use the combined arms concept to put the fire support unit in a dilemma, and have plan B ready to go. (Can you now see why it is so important to always be keeping an eye on the enemy? Gaining and maintaining visual contact is crucial!) Consider that you may be able to simply reduce his fire support unit's ability to deliver effective fire; if he is using mortars, try to knock down his aiming stakes with grenades and your own indirect fires. Once those stakes are down his fires become less accurate, and if a mortar crewman attempts to replant them, your sniper can zing him. But the mortars may be following in trace of the maneuver elements rather than being left in the base camp. If this is the case, the ambush might be planned so that it puts the mortars in a poor firing position (beneath thick canopy so that their rounds can't be fired, or in a muddy field that makes seating the base plate very difficult).

Still, a smart company commander will have a fire support plan that intends to utilize more than just his organic mortars; he might have larger, nonorganic mortars, artillery, or even close air support on call. The guerrillas' ambush, if sprung correctly (that is, extremely fast and fatal), will not be vulnerable to these weapons systems. Nevertheless, the guerrilla force must be prepared to receive pursuing fires from these weapons as it withdraws, making maximum dispersion very important.

Reserve Considerations

The enemy, when operating at strengths below battalion level (company, platoon, and squad), are somewhat unlikely to have a reserve component at night because of command and control problems inherent in night operations. Therefore, since few if any of your guerrilla operations will be conducted against units larger than a company, and given that most of your operations will be conducted at night, you will not often be forced to consider the enemy's reserve force. However, if the situation puts the enemy in a very narrow or otherwise restricted zone of action that does not lend itself well to using their entire force as maneuver elements, or if the commander's scheme of maneuver necessitates the exposing of one of his flanks or rear, the enemy may choose to keep a platoon in reserve (if the unit is a company). By carefully watching the placement of that reserve element and its security measures, the guerrillas can determine whether the reserve element itself is vulnerable to attack.

In this situation, where the reserve element is used by the enemy commander to provide security along his flank or rear, the guerrillas are looking for a gap. This might be the reserve not remaining sufficiently close to the main body during their maneuver, thus leaving it exposed, or it might become exposed when left at the line of departure to await the command to come forward.

Training in night operations is always near the front of the guerrilla leader's concerns. No guerrilla unit can ever spend too much time training in everything from individual night movement to intricate night ambushes.

About the Author

Bob Newman retired from the U.S. Marines after more than 20 years of service in assorted reconnaissance, infantry, special operations, and instructor billets. He served with Fox Company, 2nd Battalion, 4th Marines, during the Gulf War as an infantry unit leader, and spent three years as a SERE instructor at the fabled Navy SERE School in Maine. He has completed the Army Airborne Course, Navy SCUBA Course, Navy Submarine Escape Trunk Operator's Course, Marine Amphibious Reconnaissance Course, U.S. Air Force Special Operations School's Revolutionary Warfare and Dynamics of International Terrorism Courses, Navy SERE and SERE Instructor Courses, and many other formal military courses of instruction. His final tour was as a warfighting instructor at the Marine Corps Staff NCO Academy's Advanced Course aboard Camp Geiger.

The award-winning author of numerous books and thousands of magazine and newspaper articles and columns that have appeared in dozens of publications, he wrote a controversial weekly column for the Marine Corps edition of the *Navy Times* as "Dan McGrew" and still writes opinion pieces on leadership for that publication. He founded L.L. Bean's Outdoor

Discovery Program's Wilderness Survival Workshop while living in Maine. He currently lives in the Rockies where he is an editor at a publishing house, the director of the Wilderness Emergency Skills Course, and is *Soldier of Fortune*'s contributing editor for Gulf War Veterans Affairs.

Index